AF588364

# MULTIVARIATE STATISTICAL MODELING AND DATA ANALYSIS

## THEORY AND DECISION LIBRARY

General Editors: W. Leinfellner and G. Eberlein

Series A: Philosophy and Methodology of the Social Sciences

Editors: W. Leinfellner (Technical University of Vienna)
G. Eberlein (Technical University of Munich)

Series B: Mathematical and Statistical Methods

Editor: H. Skala (University of Paderborn)

Series C: Game Theory, Mathematical Programming and Mathematical Economics

Editor: S. Tijs (University of Nijmegen)

Series D: System Theory, Knowledge Engineering and Problem Solving

Editor: W. Janko (University of Vienna)

---

## SERIES B: MATHEMATICAL AND STATISTICAL METHODS

Editor: H. Skala (Paderborn)



**Scope**

The series focuses on the application of methods and ideas of logic, mathematics and statistics to the social sciences. In particular, formal treatment of social phenomena, the analysis of decision making, information theory and problems of inference will be central themes of this part of the library. Besides theoretical results, empirical investigations and the testing of theoretical models of real world problems will be subjects of interest. In addition to emphasizing interdisciplinary communication, the series will seek to support the rapid dissemination of recent results.

# MULTIVARIATE STATISTICAL MODELING AND DATA ANALYSIS

*Proceedings of the Advanced Symposium on Multivariate Modeling and Data Analysis May 15–16, 1986*

*edited by*

H. BOZDOGAN

*Department of Mathematics, University of Virginia, Charlottesville, Virginia, U.S.A.*

and

A. K. GUPTA

*Department of Mathematics and Statistics, Bowling Green State University, Bowling Green, Ohio, U.S.A.*

D. REIDEL PUBLISHING COMPANY

A MEMBER OF THE KLUWER  ACADEMIC PUBLISHERS GROUP

DORDRECHT / BOSTON / LANCASTER / TOKYO

**Library of Congress Cataloging in Publication Data**

Advanced Symposium on Multivariate Modeling and Data Analysis (1986: James Madison University).
Multivariate statistical modeling and data analysis.

(Theory and decision library. Series B, Mathematical and statistical methods)
Includes index.
1. Multivariate analysis—Congresses. I. Bozdogan, H. (Hamparsum), 1945– . II. Gupta, A. K. (Arjun K.), 1938– . III. Title. IV. Series.
QA278.A275 1986 519.5'35 87–20680
ISBN 90–277–2592–6

---

Published by D. Reidel Publishing Company,
P.O. Box 17, 3300 AA Dordrecht, Holland.

Sold and distributed in the U.S.A. and Canada
by Kluwer Academic Publishers,
101 Philip Drive, Assinippi Park, Norwell, MA 02061, U.S.A.

In all other countries, sold and distributed
by Kluwer Academic Publishers Group,
P.O. Box 322, 3300 AH Dordrecht, Holland.



Printed in The Netherlands

## CONTENTS

# PREFACE

*This volume contains the Proceedings of the Advanced Symposium on Multivariate Modeling and Data Analysis held at the 64th Annual Meeting of the Virginia Academy of Sciences (VAS)--American Statistical Association's Virginia Chapter at James Madison University in Harrisonburg, Virginia during May 15-16, 1986.*

*This symposium was sponsored by financial support from the Center for Advanced Studies at the University of Virginia to promote new and modern information-theoretic statistical modeling procedures and to blend these new techniques within the classical theory. Multivariate statistical analysis has come a long way and currently it is in an evolutionary stage in the era of high-speed computation and computer technology. The Advanced Symposium was the first to address the new innovative approaches in multivariate analysis to develop modern analytical and yet practical procedures to meet the needs of researchers and the societal need of statistics.*

*Papers presented at the Symposium by emminent researchers in the field were geared not just for specialists in statistics, but an attempt has been made to achieve a well balanced and uniform coverage of different areas in multivariate modeling and data analysis. The areas covered included topics in the analysis of repeated measurements, cluster analysis, discriminant analysis, canonical correlations, distribution theory and testing, bivariate density estimation, factor analysis, principle component analysis, multidimensional scaling, multivariate linear models, nonparametric regression, etc.*

*The program organizing committee included Hamparsum Bozdogan as the program host and chairman, University of Virginia; Arjun K. Gupta as co-chairman, Bowling Green State University; Donald E. Ramirez, University of Virginia; and Ms. Julie Riddleberger as the secretary who efficiently handled all the correspondence, typing the abstracts of the papers, and the announcements for the Symposium.*

*We regret that it was not possible to include the paper, "Nonparametric Regression for Censored Survival Data," by Professor John Van Ryzin (who spoke at the Symposium) due to his untimely death. We gratefully acknowledge his contribution and encouragement throughout the organization of the Symposium.*

*Finally, we extend our thanks to Dean W. Dexter Whitehead, Director of the Center for Advanced Studies at the University of Virginia, for the financial support; to Golde Holtzman of the Department of Statistics at Virginia Tech for his help in the planning of the program; and to several referees who reviewed the papers. We also thank all the contributors to this volume and the Symposium.* *We hope that these contributions will stimulate more research and help further the advancement of the field of multivariate statistical modeling and data analysis.*

*Charlottesville, May 1987* *H. Bozdogan*

*Bowling Green, May 1987* *A. K. Gupta*

Taskin Atilgan and Tom Leonard

# ON THE APPLICATION OF AIC TO BIVARIATE DENSITY ESTIMATION, NONPARAMETRIC REGRESSION AND DISCRIMINATION

## ABSTRACT

Some simple data analytic procedures are available for bivariate nonparametric density estimation. If we use a linear approximation of specified basis functions then the coefficients can be estimated by the EM algorithm, and the number of terms judged by Akaike's information criterion. The method also yields readily compatible approaches to nonparametric regression and logistic discrimination. Tukey's energy consumption data and a psychological test for 25 normal and 25 psychotic patients are re-analyzed and the current methodology compared with previous procedures. The procedures offer many possible applications in the biomedical area, which are discussed in Sections 5 and 6, e.g. it is possible to analyze noisy data sets in situations where structured regression techniques would typically fail.

KEYWORDS: AIC, Roughness parameter; Bias; Variance; Tradeoff.

## 1. INTRODUCTION

Let $\{(x_i, y_i); i=1, \cdots, n\}$ denote a bivariate random sample of n observations from a population with an unknown density $f(x, y)$. Consider linear approximations of the form

$$f_m(x, y) = \sum_{i=1}^{m} \theta_i \, \psi_i(x, y) \qquad \left[\theta_i \geq 0 \text{ for } i=1, \cdots, m; \quad \sum_{i=1}^{m} \theta_i = 1\right] \tag{1.1}$$

for $f(x, y)$, where the $\psi_i$ are bivariate densities of known functional form and m is the dimension of the approximation. Assume that each $\psi_i$ contains the same number of unknown parameters. For any dimension m it is possible to obtain estimates $\hat{\theta}_1, \cdots, \hat{\theta}_m$ for $\theta_1, \cdots, \theta_m$ and $\hat{\psi}_1, \cdots, \hat{\psi}_m$ for $\psi_1, \cdots, \psi_m$ using maximum likelihood. Let

$$\hat{L}_m = \sum_{j=1}^{n} \log \left\{\sum_{i=1}^{m} \hat{\theta}_i \hat{\psi}_i(x_j, y_j)\right\} \tag{1.2}$$

denote the corresponding maximized loglikelihood. Then Atilgan (1984)



*H. Bozdogan and A. K. Gupta (eds.), Multivariate Statistical Modeling and Data Analysis, 1–16.*

demonstrates the asymptotic behavior

$$E\left\{\frac{1}{n}\hat{L}_m\right\} = \frac{m-1}{2n} - I\left[f, f_m^o\right] + NE + 0(n^{-2}) \qquad (n \to \infty) \tag{1.3}$$

where

(a) (m–1)/2n is a "variance" term which increases with dimension

(b) $$f_m^o = E\left[\hat{f}_m \mid f\right] \tag{1.4}$$

(c) $$I\left[f, f_m^o\right] = \int\int f(x, y) \log \left[f(x, y)/f_m^o(x, y)\right] dx\, dy \tag{1.5}$$

is the Kullback-Leibler (1951) information distance between the sampling expectation $f_m^o$ and the true density f. This plays a similar role to "bias" for single parameter problems. Whenever $f_1, f_2, \cdots$ from (1.1) represent a nested family of approximations, $I[f, f_m^o]$ will decrease as m increases.

(d) $$NE = \text{Negative Entropy} = \int\int f(x, y) \log f(x, y)\, dx\, dy\ . \tag{1.6}$$

Atilgan (1984) showed that

$$-E\left\{\frac{1}{n}\hat{L}_m - (1+\alpha)\frac{(m-1)}{2n}\right\} = I\left[f, f_m^o\right] + \frac{\alpha(m-1)}{2n} + NE + 0(n^{-2})\ , \tag{1.7}$$

where $\alpha > 0$ is called a roughness parameter and controls the tradeoff between the "bias" term $I[f, f_m^o]$ and the "variance" term (m–1)/2n. For a given sample size n, this motivates us to check m by maximizing the penalized log-likelihood

$$GIC = \frac{1}{n}\hat{L}_m - (1+\alpha)\frac{m-1}{2n} \tag{1.8}$$

which provides a general information criterion.

There are various possible choices for $\alpha$. Atilgan (1983) demonstrated via computer simulation that the choice $\alpha=1$ is often reasonable, leading to maximization of Akaike's Information Criterion

$$\text{AIC} = \hat{L}_m - m \;. \tag{1.9}$$

The roughness parameter $\alpha$ controls the fidelity of the approximation to the data relative to the smoothness of the linear approximation. Smaller $\alpha$ values will give better fidelity to the data, but a rougher estimate for f, than will larger values of $\alpha$. The value $\alpha$=1 can be interpreted as giving equal weight to both the "bias" and the "variance" terms. The reason we use (1.9) rather than the standard form of AIC $= -2\hat{L}_m + 2m$, is that in (1.9) the extention of maximum likelihood method from estimation to model selection is more transparent; maximize $L_m$ to obtain estimates of the parameters of the model and maximize $\hat{L}_m - m$ over m to select the model.

## 2. BIVARIATE DENSITY ESTIMATION AND NON PARAMETRIC REGRESSION

Consider the two possible choices

(A)

$$\psi_i(x, y) = \begin{cases} 1 & \text{if } (x, y) \varepsilon A_i \\ \\ 0 & \text{otherwise} \end{cases} \qquad (i=1, \cdots, m) \tag{2.1}$$

where $A_1, \cdots, A_m$ comprise a partition of $B \subseteq R^2$ (this produces a histogram estimate for f concentrated on B)

and

(B) $\psi_i(x, y) =$ a bivariate normal density with mean vector

$$\underset{\sim}{\mu_i} = \left(\mu_{x_i}, \mu_{y_i}\right)^T \qquad (i=1, \cdots, m) \tag{2.2}$$

and covariance matrix

$$\Sigma = \begin{bmatrix} \sigma_1^2 & \rho\sigma_1\sigma_2 \\ \\ \rho\sigma_1\sigma_2 & \sigma_2^2 \end{bmatrix} \tag{2.3}$$

of basis functions for the linear approximation in (1.1).

In either case we can estimate f(x, y) via the maximum likelihood/AIC procedure outlined in Section 1. If $\hat{f}_m(x, y)$ denotes our estimate then the regression of y on x may be estimated by

$$r(x) = \int_{R^1} y\, \hat{f}_m(x, y)\, dy/\hat{f}_1(x) \tag{2.4}$$

where

$$\hat{f}_1(x) = \int_{R^1} \hat{f}_m(x, y)\, dy\,. \tag{2.5}$$

The histogram case described above in (A) is particularly useful as the interpretation in (2.4) and (2.5) may be replaced by summations.

## 3. THE BIVARIATE HISTOGRAM SITUATION

Suppose here that all n observations $(x_i, y_i)$ are concentrated within a rectangle $B = (a_o, a_1)\times(b_o, b_1)$. Then bivariate histogram estimator with $m = k\times k$ equal rectangular meshes is given by

$$f_m(x, y) = \sum_{i=1}^{k} \sum_{j=1}^{k} P_{ij}\, I_{B_{ij}}(x, y)/\omega_x\omega_y \tag{3.1}$$

where

$$B_{ij} = (a_o + (i-1)\,\omega_x,\ a_o + i\omega_x) \times (b_o + (j-1)\,\omega_y,\ b_o + j\omega_y) \tag{3.2}$$

with $\omega_x = (a_1 - a_o)/k$, $\omega_y = (b_1 - b)/k$ and $P_{ij}$ denoting the proportion of observations falling in $B_{ij}$.

In this case m should be chosen to maximize

$$\begin{aligned} AIC &= \hat{L}_m - m \\ &= C + \sum_{i=1}^{k} \sum_{j=1}^{k} P_{ij} \log P_{ij} - n \log m - m \end{aligned} \tag{3.3}$$

where C does not depend upon m.

Then the regression function in (2.4) reduces to

$$r(x) = \sum_{i=1}^{k} P_{ij} \left[ b_o + \left( j - \frac{1}{2} \right) \omega_y \right] / \sum_{j=1}^{k} P_{ij} \tag{3.4}$$

$$\text{for } x \,\varepsilon (a_o + (i-1)\,\omega_x,\ a_o + i\omega_x), \qquad (i=1, \cdots, k)\ .$$

This yields a piecewise estimated regression function where the number of vertices is chosen according to AIC.

## 4. BIVARIATE NORMAL BASIS FUNCTIONS

Suppose now that for $i=1, \cdots, m = k^2$, $\psi_i$ is the bivariate normal density identified as option (B) in Section 2. We moreover, fix the mean vectors $\underset{\sim}{\mu}_i$ to lie on a grid by setting $\underset{\sim}{\mu}_i = \underset{\sim}{\hat{\mu}}_i = (\hat{\mu}_{x_i}, \hat{\mu}_{y_i})$ where $\underset{\sim}{\hat{\mu}}_i$ is a distinct point of the array $\{(\eta_g^{(1)}, \eta_h^{(2)});\ g=1, \cdots, k;\ h=1, \cdots, k\}$ with

$$\eta_g^{(1)} = x_{min} + \frac{g}{k+1}(x_{max} - x_{min}) \qquad (g=1, \cdots, k) \tag{4.1}$$

and

$$\eta_h^{(2)} = y_{min} + \frac{h}{k+1}(y_{max} - y_{min}), \qquad (h=1, \cdots, k)\ . \tag{4.2}$$

The EM algorithm (see Dempster *et al.*, 1978) can now be used to estimate the common covariance matrix $\Sigma$ in (2.3) together with the mixing probabilities $\theta_1, \cdots, \theta_m$. The steps of the EM algorithm are:

$$\text{E Step:} \quad \Pi_i(x_i, y_j) = \frac{\theta_i^{(P)} \psi_i^{(P)}(x_j, y_j)}{\sum_{d=1}^{k^2} \theta_d^{(P)} \psi_d^{(P)}(x_j, y_j)} \qquad \begin{array}{l} i=1, \cdots, k^2 \\ j=1, \cdots, n \end{array} \tag{4.3}$$

where $\psi_i^{(P)}$ denotes a bivariate normal density with mean vector $\underset{\sim}{\hat{\mu}}_i$ and covariance matrix

$$\Sigma^{(P)} = \begin{bmatrix} \sigma_1^{2(P)} & \rho^{(P)}\sigma_1^{(P)}\sigma_2^{(P)} \\ \rho^{(P)}\sigma_1^{(P)}\sigma_2^{(P)} & \sigma_2^{2(P)} \end{bmatrix} \tag{4.4}$$

M Step: $$\theta_i^{(P+1)} = \frac{1}{n}\sum_{j=1}^{n} \Pi_i(x_j, y_j) \quad (i=1, \cdots, k^2) \tag{4.5}$$

$$\sigma_1^{2(P+1)} = \frac{1}{n}\sum_{j=1}^{n}\sum_{i=1}^{k^2} \Pi_i \left(x_j, y_j\right) \left(x_j - \mu_{x_i}\right)^2 \tag{4.6}$$

$$\sigma_2^{2(P+1)} = \frac{1}{n}\sum_{j=1}^{n}\sum_{i=1}^{k^2} \Pi_i \left(x_j, y_j\right) \left(y_j - \mu_{y_i}\right)^2 \tag{4.7}$$

$$\rho^{(P+1)} = \sum_{j=1}^{n}\sum_{i=1}^{k^2} \Pi_i \left(x_j, y_j\right) \left(x_j - \mu_{x_i}\right) \left(y_j - \mu_{y_i}\right) / \sigma_1^{(P)}\sigma_2^{(P)} \tag{4.8}$$

where (P) and (P+1) represent the Pth and (P+1)th iterations.

It is possible to choose the dimension m, by referring to AIC in (1.9), but with $m+2$ unknown parameters. Let

$$\hat{f}_m(x, y) = \sum_{i=1}^{m} \hat{\theta}_i \hat{\psi}_i(x, y) \tag{4.9}$$

denote our final estimate for f where $\hat{\psi}_i$ has mean vector $(\hat{\mu}_{x_i}, \hat{\mu}_{y_i})^T$ and estimated variances $\hat{\sigma}_1^2$ and $\hat{\sigma}_2^2$ and correlation $\hat{\rho}$.

The estimated regression function of y on x is now

$$r(x) = \sum_{j=1}^{m} w_i \left\{ \hat{\mu}_{y_i} + \hat{\rho}\, \frac{\hat{\sigma}_2^2}{\hat{\sigma}_1} \left(x - \hat{\mu}_{x_i}\right) \right\} \tag{4.10}$$

where

$$w_i = \hat{\theta}_i \frac{1}{\sqrt{2\pi}\hat{\sigma}_1} \exp\left\{ -\left(x - \hat{\mu}_{x_i}\right)^2 / 2\hat{\sigma}_1^2 \right\} / \hat{f}_1(x) \tag{4.11}$$

with

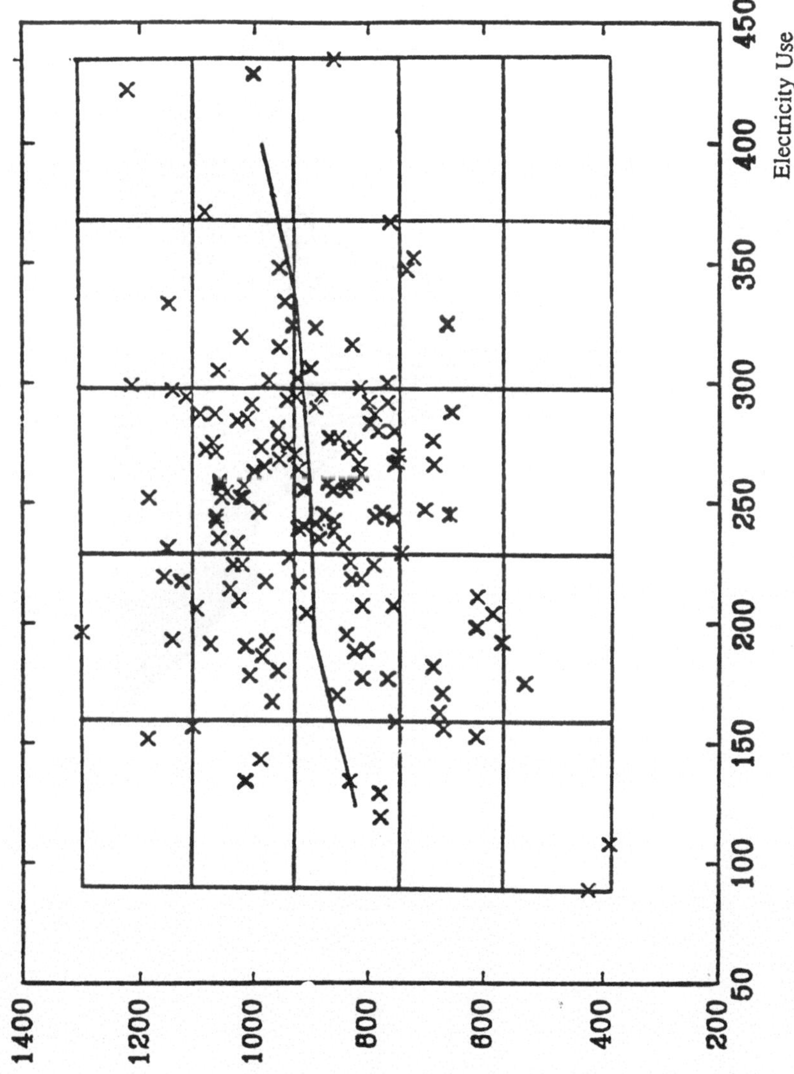


Figure 1. Non-parametric regression with bivariate histogram, m=25; Tukey's data.

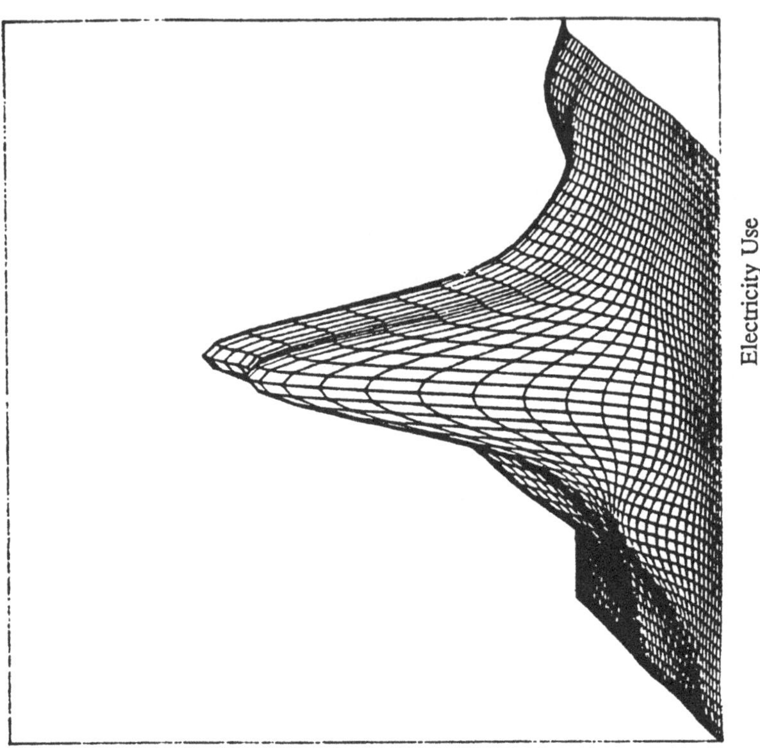


Figure 2. Bivariate density estimate using bivariate normal basis, m=25; Tukey's data.

$$\hat{f}_1(x) = \sum_{i=1}^{m} \hat{\theta}_i \frac{1}{\sqrt{2\pi\hat{\sigma}_1}} \exp\left\{-\frac{1}{2\hat{\sigma}_1^2}\left[x-\hat{\mu}_{x_i}\right]^2\right\}. \tag{4.12}$$

Therefore instead of the more common conditional/mean regression, (4.10) weights m separate linear regressions where the weights $w_i$ themselves depend upon x. This provides a smooth nonparametric estimate of the regression function.

## 5. APPLICATION TO TUKEY'S DATA

Consider the data more fully described by Tukey (1977, p. 267) and relating to the regression of gas consumption upon electricity consumption for 152 townhouses in Twin Rivers, New Jersey.

For our histogram based procedure AIC attained a maximum at $m=25$ i.e., $k=5$. The corresponding regression function is described in Figure 1.

For our bivariate normal basis functions, AIC again suggested $m=25$. The bivariate density estimate is depicted in Figure 2, and the corresponding contour plot and regression function are in figures 3 and 4.

Whilst the data are extremely noisy our procedure enables us to calculate valid regression functions which demonstrate a positive relationship between gas and electricity consumption which is not however totally linear. It would be difficult to employ the standard techniques of the linear statistical model to draw similar conclusions. The estimate in Figure 2 is clearly bimodal, probably an effect of gas use. This may be due to some inhomogeneity in the 152 townhouses.

Similar techniques may be applied to a variety of biomedical problems where the noisiness of the data and inherent non-linearity make it difficult to model a parametric regression function by more standard methods. Many biomedical data sets, involving response levels, and symptom levels do not possess much *predictive* content in the sense that it is difficult to precisely predict a response level, given a symptom level. They may however still possess substantial *probabilistic* content in the sense that probabilities for the response levels vary substantially with the symptom level. By modeling the full joint distribution of the response and the symptom, without too many constrictive assumptions, it is possible to quantify this probabilistic content in a meaningful way, in situations where standard regression techniques may well fail.

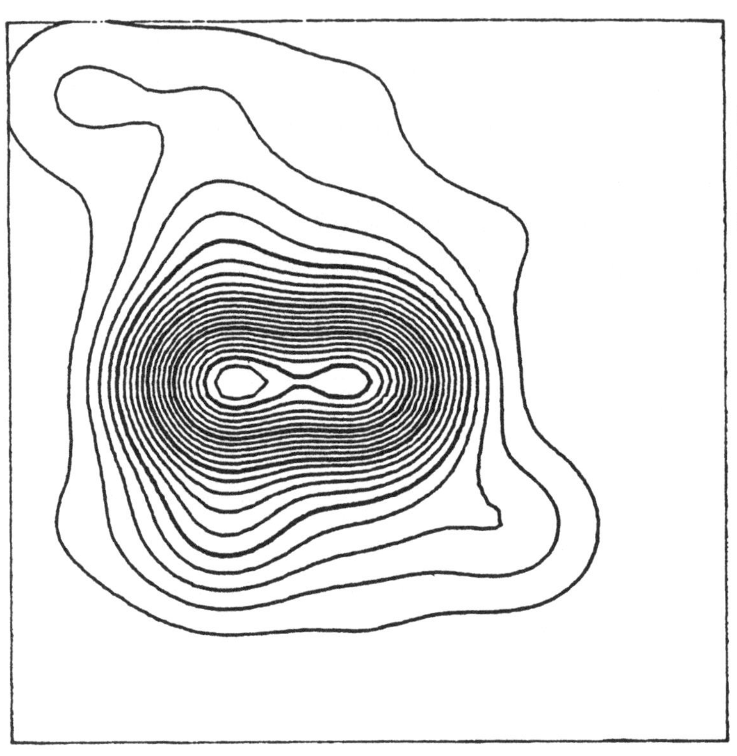


Figure 3. Contours of the density estimate using bivariate normal basis, m=25; Tukey's data.

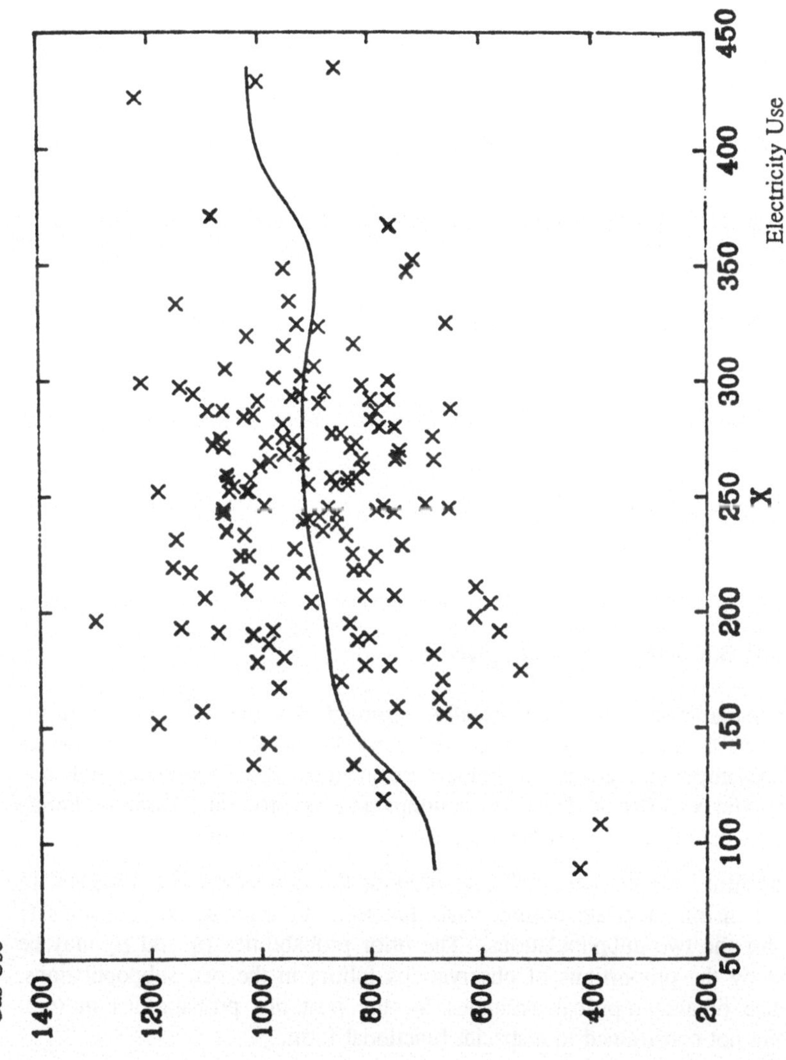


**Figure 4.** Non-parametric regression functions using bivariate normal basis, m= 25; Tukey's data

## 6. NONPARAMETRIC DISCRIMINANT ANALYSIS

Let $H_1, \cdots, H_k$ be k populations and $\underset{\sim}{x} = (x_1, \cdots, x_d)$ be a d-dimensional random vector. Let $f_j(\underset{\sim}{x})$ denote the p.d.f. of $\underset{\sim}{x}$ under population $H_j$, and let $p_j$ denote the prior probability of $\underset{\sim}{x}$ under population $H_j$, $(j=1, \cdots, k)$.

Under standard discrimination techniques an observation $\underset{\sim}{x}$ is allocated to population $H_s$ if

$$p(H_s \mid \underset{\sim}{x}) > p(H_j \mid \underset{\sim}{x}) \qquad (j \neq s;\ j=1, \cdots, k) \tag{6.1}$$

where

$$p(H_j \mid \underset{\sim}{x}) = \frac{p_j f_j(\underset{\sim}{x})}{\sum_{i=1}^{k} p_i f_i(\underset{\sim}{x})} \tag{6.2}$$

is the posterior probability of $H_j$, given $\underset{\sim}{x}$.

Special parametric forms are usually assumed for the $f_j$. For example, Fisher (1938), Smith (1947), and Anderson (1972) show that multivariate normal assumptions lead to either linear or quadratic discrimination. Habema, Hermans and Van Broek (1974) use a nonparametric procedure based on kernel methods, where there is a problem determining the band width.

We concentrate on the case k=2 and propose the nonparametric procedure of Section 4, using bivariate normal basis functions to estimate the densities $f_1$ and $f_2$ for the two subpopulations. The prior probabilities $p_1$ and $p_2$ may be replaced by the proportions of observations falling in the two subpopulations. We hence obtain empirical estimates for the posterior probabilities in (6.2) which are not constrained to a special functional form.

As an example, Smith (1947) and Anderson (1975) consider measurements for a psychological test on a group of 25 normal and 25 psychotic patients with measurements of "size", x, and "shape" y for each patient.

In this case AIC for the method in Section 4, gave m=4 as the optimal number of bivariate normal terms for both the normal and the psychotic patients. The

Table I Empirical posterior probabilities for "normals" and "psychotics"

| "Normals" | | | "Psychotics" | | |
|---|---|---|---|---|---|
| x "size" | y "shape" | Pr(Psychotic\|x,y) | x "size" | y "shape" | Pr(Psychotic\|x,y) |
| 22.00 | 6.00 | 0.01 | 24.00 | 38.00 | 0.99 |
| 20.00 | 14.00 | 0.03 | 19.00 | 36.00 | 0.88 |
| 23.00 | 9.00 | 0.01 | 11.00 | 43.00 | 0.99 |
| 23.00 | 1.00 | 0.01 | 6.00 | 60.00 | 1.00 |
| 17.00 | 8.00 | 0.14 | 9.00 | 32.00 | 0.99 |
| 24.00 | 9.00 | 0.02 | 10.00 | 17.00 | 0.98 |
| 23.00 | 13.00 | 0.06 | 3.00 | 17.00 | 1.00 |
| 18.00 | 18.00 | 0.16 | 15.00 | 56.00 | 1.00 |
| 22.00 | 16.00 | 0.13 | 14.00 | 43.00 | 0.99 |
| 19.00 | 18.00 | 0.11 | 20.00 | 8.00 | 0.04 |
| 20.00 | 17.00 | 0.08 | 8.00 | 46.00 | 1.00 |
| 20.00 | 31.00 | 0.42 | 20.00 | 62.00 | 1.00 |
| 21.00 | 9.00 | 0.01 | 14.00 | 36.00 | 0.99 |
| 13.00 | 13.00 | 0.49 | 3.00 | 12.00 | 1.00 |
| 20.00 | 14.00 | 0.03 | 10.00 | 51.00 | 1.00 |
| 19.00 | 15.00 | 0.06 | 22.00 | 22.00 | 0.14 |
| 20.00 | 11.00 | 0.03 | 11.00 | 30.00 | 0.99 |
| 18.00 | 17.00 | 0.14 | 6.00 | 30.00 | 0.99 |
| 20.00 | 7.00 | 0.05 | 20.00 | 61.00 | 1.00 |
| 23.00 | 6.00 | 0.00 | 20.00 | 43.00 | 0.99 |
| 23.00 | 23.00 | 0.12 | 15.00 | 48.00 | 0.99 |
| 25.00 | 9.00 | 0.05 | 5.00 | 53.00 | 1.00 |
| 23.00 | 5.00 | 0.00 | 10.00 | 43.00 | 0.99 |
| 21.00 | 12.00 | 0.02 | 13.00 | 19.00 | 0.71 |
| 23.00 | 7.00 | 0.00 | 12.00 | 4.00 | 0.99 |

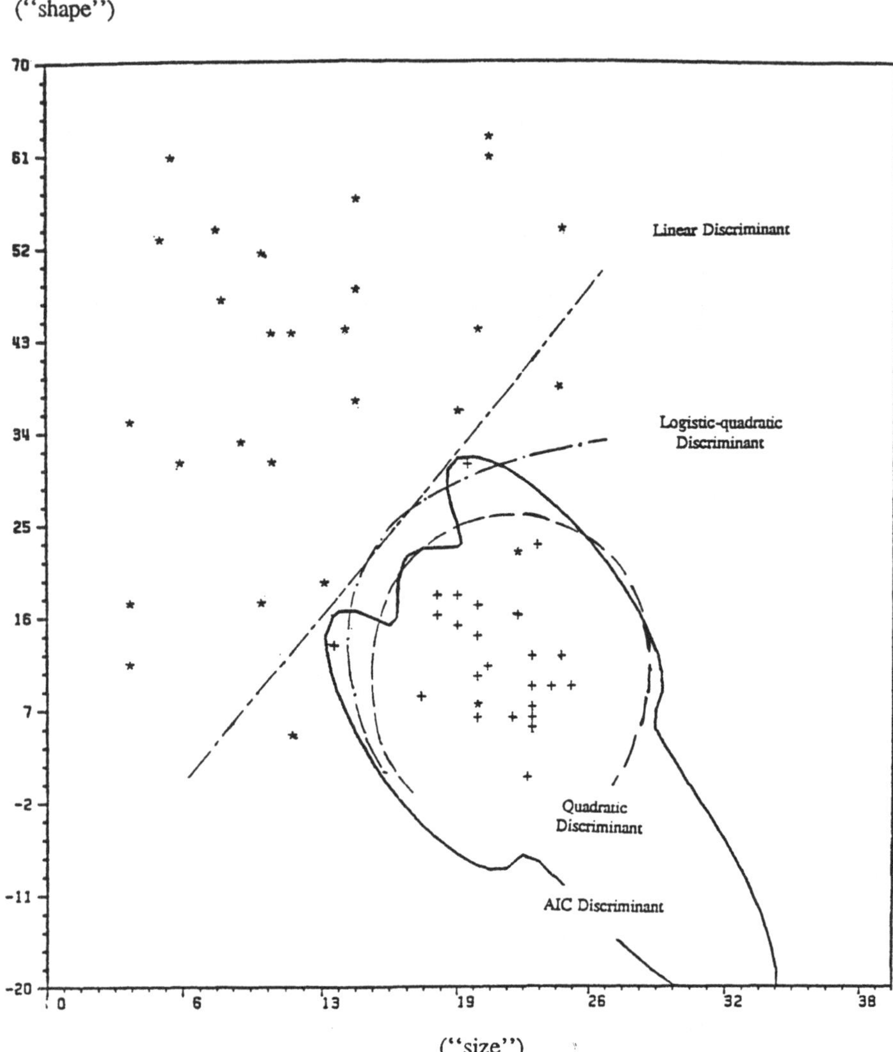


Figure 5. Discrimination of "normals" (+) and "psychotics" (*)

contour showing $f_1 = f_2$ ($f_1$ is the density for normals and $f_2$ is for psychotics) is depicted in Figure 5. Figure 5 also shows linear, quadratic and logistic quadratic discriminants. We see from Figure 5 that while the number of misclassified observations for linear, quadratic and logistic-quadratic discriminants is 4 it is only 2 for the nonparametric discriminant. Table I presents the ''posterior'' probabilities of a person being psychotic given the two measurements, ''size'' and ''shape'', for psychotic patients, and also for normal patients.

## 7. CONCLUSION

We suggest that the simple data analytic techniques discussed in Sections 5 and 6 should be useful for the analysis of many biomedical data sets where the noisiness of the data set and non-linearity of appropriate models make it difficult to apply more standard regression and discrimination procedures.

Taskin Atilgan*
Department of Statistics
Texas A&M University
College Station
Texas 77843-3143

Tom Leonard
Department of Statistics
University of Wisconsin
Madison, WI 53706

* Presently at the AT&T Bell Laboratories, Murray Hill, New Jersey, 07974. The Computing time provided by the Texas A&M University and University of Wisconsin-Madison is greatfully acknowledged.

## REFERENCES


Akaike, H. (1974). 'A New Look at the Statistical Model Identification.' I.E.E.E. Transactions on Automatic Control, **AC-19, 716-712.**

Anderson, J. R. (1972). 'Separate sample logistic discrimination.' Biometrika **59, 19-35.**

Anderson, J. R. (1975). 'Quadratic logistic discrimination.' Biometrika **62, 149-154.**

Atilgan, T. (1983). 'Parameter Parsimony Model Selection, and Smooth Density Estimation,' Ph.D. Thesis Department of Statistics, University of Wisconsin-Madison.

Atilgan, T. (1984). 'Determination of the Dimension of an Approximation for Fitting a Given Data Set.' Tech. Rep. No. 741 Department of Statistics, University of Wisconsin, Madison.

Dempster, A. P., Laird, N. H. and Rubin, D. B. (1978). 'Maximum Likelihood Incomplete Data via the EM Algorithm.' J.R.S.S., Series B, **39:1-38.**

Fisher, R. A. (1938). 'The statistical utilization of multiple measurements.' Annals of Eugenics **8:376-386.**

Habema, J. D. F., Hermans, J. and Van den Broek, K. (1974). 'A Stepwise Discriminant Analysis Program Using Density Estimation.' In: COMPSTAT 1974, Proceedings in Computational Statistics. (G. Bruckman, ed.) Physica Verlag, Wien.

Kullback, S. and Leibler, R. A. (1951). 'On information and sufficiency.' Annals of Mathematical Statistics, **22**, 79-86.

Smith, C. A. B. (1947). 'Some examples of Discrimination.' Annals of Eugenics, **13:272-282.**

Tukey, J. W. (1977). 'Exploratory Data Analysis.' Addison-Wesley.

H.H. BOCK

# ON THE INTERFACE BETWEEN CLUSTER ANALYSIS, PRINCIPAL COMPONENT ANALYSIS, AND MULTIDIMENSIONAL SCALING

Abstract: This paper shows how methods of cluster analysis, principal component analysis, and multidimensional scaling may be combined in order to obtain an optimal fit between a classification underlying some set of objects 1,...,n and its visual representation in a low-dimensional euclidean space $\mathbb{R}^s$. We propose several clustering criteria and corresponding k-means-like algorithms which are based either on a probabilistic model or on geometrical considerations leading to matrix approximation problems. In particular, a MDS-clustering strategy is presented for displaying not only the n objects using their pairwise dissimilarities, but also the detected clusters and their average distances.

## 1. INTRODUCTION

Given a set of sampled observation vectors $x_1,\ldots,x_n \in \mathbb{R}^p$, partition-type clustering methods try to subdivide $\{x_1,\ldots,x_n\}$ respectively the corresponding set of objects $\mathcal{O} = \{1,\ldots,n\}$ into a number m of classes $C_1,\ldots,C_m$ such that these classes are as homogeneous as possible and the resulting partition $\mathcal{C} = (C_1,\ldots,C_m)$ is best adapted to the data set. The mathematical and statistical formulation of this problem may be given in a number of different ways by using, alternatively, probabilistic models, graph-theoretical terms, or optimization criteria and algorithms. This latter formulation will be adopted in this paper, too: By defining and minimizing a well-chosen clustering criterion we will construct, or approximate, an optimal resp. sub-optimal classification of the data set. As a rule, we shall assume the class number m to be known beforehand. In cases where this number is unknown to the data analysist, he may apply, e.g., the proposed algorithms iteratively for a series of values m = 2,3,... .

Another well-known method for extracting the essential information from a given set of data proceeds by displaying the



*H. Bozdogan and A. K. Gupta (eds.), Multivariate Statistical Modeling and Data Analysis, 17–34.*

different objects 1,...,n by n points $x_1^*,\ldots,x_n^*$ in some low-dimensional (in most cases: two-dimensional) picture which summarizes in a simplified way, but preserves as much as possible, the original full information contained in the whole data set. In particular, if the dimension p of the data vectors (i.e., the number of observed variables) is large, a projection onto an optimally suited hyperplane of dimension s=2 may be appropriate - this is the procedure adopted in principal component analysis.

The same geometrical viewpoint is underlying the well-known methods of multidimensional scaling (MDS): In this case, the data analysist has at its disposal, instead of n data points, only some knowledge on the mutual similarity structure of the given set of objects 1,...,n, expressed by an $n \times n$ matrix $D = (d_{k\ell})$ where $d_{k\ell}$ denotes the dissimilarity between the pair of objects $k,\ell \in \mathcal{O}$. (We will not discuss the problems concerned with the definition, computation, or empirical evaluation of dissimilarity indices.) From this information, MDS methods compute a low-dimensional representation of the n objects by n points $x_1^*,\ldots,x_n^* \in \mathbb{R}^s$ such that the euclidean distances $\|x_k - x_\ell\|$ of these points approximate, as best as possible, the given dissimilarity values $d_{k\ell}$. (Actually, we will use the scalar product version of MDS explained in Section 3.)

In the framework of cluster analysis (CA) it is generally expected that a possible 'natural' clustering of the objects will be evident from a look at the points $x_1^*,\ldots,x_n^*$ obtained from PCA or MDS, e.g., from the two-dimensional visualization. However, in principle neither principal component analysis nor multidimensional scaling are designed for this purpose, i.e. for choosing $x_1^*,\ldots,x_n^*$ so carefully that an existing clustering can be optimally detected from the display. Thus, the usual PCA or MDS methods risk to miss an underlying classification of the data even if the dimension s is chosen large enough.

In this paper we show how this drawback can be remedied: By a suitable definition of an optimality criterion $g(\mathcal{C},X)$ and an iterative k-means like algorithm, we are able to characterize, and compute, an optimal euclidean representation of the objects and, *simultaneously*, an optimal clustering $\mathcal{C}$ such that each of them is best suited to the other one. More generally, in Section 2 we shall propose and review a series of clustering methods which take account of a linear structure of clusters or which are related otherwise to PCA. In most cases we shall adopt a geometrical point of view, but in the Sections 2.2, 2.3 a probabilistic clustering model will be presented, too, where maximum likelihood estimation leads directly to the geometrically motivated clustering methods. Section 3 is devoted to MDS clustering methods starting from a dissimilarity matrix D: We investigate how MDS representations can allow for a given classification $\mathcal{C}$ of the objects $1,\dots,n$ and we propose an iterative MDS clustering method for obtaining, additionally, an optimum choice of $\mathcal{C}$.

## 2. PRINCIPAL COMPONENT CLUSTERING: SEVERAL ALTERNATIVES

Suppose that we are given n data points $x_1,\dots,x_n \in \mathbb{R}^p$ representing n objects $1,\dots,n$. Throughout the paper, we shall denote by $\mathcal{C} = (C_1,\dots,C_m)$ an arbitrary m-partition of $\mathcal{O} = \{1,\dots,n\}$, by $n_i := |C_i|$ the class sizes, and by

$$\bar{x} := \frac{1}{n}\sum_{k=1}^{n} x_k \qquad \bar{x}_{C_i} := \frac{1}{n_i}\sum_{k\in C_i} x_k$$

the mean vector resp. the m class means. Furthermore

$$S := \sum_{k=1}^{n} (x_k-\bar{x})(x_k-\bar{x})' \qquad B(\mathcal{C}) := \sum_{i=1}^{m} n_i(\bar{x}_{C_i}-\bar{x})(\bar{x}_{C_i}-\bar{x})'$$
$$W(\mathcal{C}) := \sum_{i=1}^{m} W(C_i) \qquad W(C_i) := \sum_{k\in C_i} (x_k-\bar{x}_{C_i})\,(x_k-\bar{x}_{C_i})' \tag{2.1}$$

will denote the p×p scatter matrices of the sample (total, between the classes, in the classes of $\mathcal{C}$, in the class $C_i$).

### 2.1 Ordinary Principal Component Analysis (PCA)

Ordinary PCA solves the following problem: For a given (small) dimension $s < p$ we look for an s-dimensional hyperplane $H \subset \mathbb{R}^p$ and n points $y_1,\dots,y_n \in H$ such that the mean squared error

$$g_0(H;y_1,\dots,y_n) := \frac{1}{n}\sum_{k=1}^{n} \|x_k-y_k\|^2 \;\to\; \min_{H,Y} \tag{2.2}$$

will be minimized. In the sequel a hyperplane H will be denoted, without further notice, by $H = a + [v_1,\dots,v_s]$ where $a \in \mathbb{R}^p$ is an arbitrary point of H and $v_1,\dots,v_s \in \mathbb{R}^p$ is supposed to be a set of orthogonal unit vectors spanning the subspace $[v_1,\dots,v_s]$ of $\mathbb{R}^p$. Moreover we shall use the matrix $V = (v_1,\dots,v_s) \in \mathbb{R}^{p\times s}$ with $V'V = I_s$ and denote by $\overline{V} = (v_{s+1},\dots,v_p) \in \mathbb{R}^{p\times(p-s)}$ a complementary matrix such that $v_{s+1},\dots,v_p$ span the orthogonal complement of $[v_1,\dots,v_s]$.

Using $a = \overline{x}$ and $y_k = P_H(x_k)$, the projection of $x_k$ onto H, (2.2) is equivalent to

$$g_1(H) := \sum \|x_k - P_H(x_k)\|^2 = tr(\overline{V}'S\overline{V}) \rightarrow \min_{\overline{V}}. \tag{2.3}$$

The solution $H^*$ of this problem is well-known (PEARSON 1901, RAO 1964, OKAMOTO 1968, BOCK 1974, §25) and involves the eigenvalues $\lambda_1 \geq \lambda_2 \geq \dots \geq \lambda_p = 0$ of the scatter matrix S and its corresponding orthonormalized eigenvectors $v_1,\dots,v_p \in \mathbb{R}^p$: Then $H^* = \overline{x} + [v_1,\dots,v_s]$ is the hyperplane spanned by the s 'largest' eigenvectors $v_1,\dots,v_s$ of S, the approximating points are given by

$$y_k = \overline{x} + \sum_{i=1}^{s} v_i\, x^*_{ki} \in H \qquad \text{with} \quad x^*_{ki} := v_i'(x_k - \overline{x}), \tag{2.4}$$

and the principal components $x^*_{ki}$ provide an optimal representation of the objects $1,\dots,n$ in $\mathbb{R}^s$ by the points

$$x^*_k = (x^*_{k1},\dots,x^*_{ks})' = V'(x_k - \overline{x}) \in \mathbb{R}^s \qquad k = 1,\dots,n. \tag{2.5}$$

The notation introduced in this Section will be used with some modifications throughout this paper.

## 2.2 PCA-based clustering with class-specific hyperplanes

Remember that the usual k-means clustering algorithm is designed for minimizing the variance criterion (mean squared error criterion):

$$g_2(\mathcal{C}) := \sum_{i=1}^{m} \sum_{k\in C_i} \|x_k - \overline{x}_{C_i}\|^2 = tr(W(\mathcal{C})) \rightarrow \min_{\mathcal{C}} \tag{2.6}$$

over all m-partitions $\mathcal{C}$ (BOCK 1974, §15). Evidently it is appropriate only for detecting ball-like classes centered at some *point* of $\mathbb{R}^p$ estimated by $\bar{x}_{C_i}$. — In contrast, in this Section, we consider m-partitions $\mathcal{C} = (C_1,\ldots,C_m)$ of $\mathcal{O}$ where each class shows some linear structure and is characterized by an (unknown) class-specific s-dimensional *hyperplane* $H_i = a_i + [v_{i1},\ldots,v_{is}] \in \mathbb{R}^p$. In fact, denoting by

$$d(x,H_i) := \min_{y \in H_i} \{ \|x-y\|^2 \} = \|x - P_{H_i}(x)\|^2$$

the squared minimum distance between a given point $x \in \mathbb{R}^p$ and the hyperplane $H_i$ we want to minimize the clustering criterion (approximation error)

$$g(\mathcal{C},\mathcal{H}) := \sum_{i=1}^{m} \sum_{k \in C_i} d(x,H_i) \rightarrow \min_{\mathcal{C},\mathcal{H}} \tag{2.7}$$

over all m-partitions $\mathcal{C}$ *and* all systems $\mathcal{H} = \{H_1,\ldots,H_m\}$ of s-dimensional hyperplanes.

Apart from typical classification problems with 'elongated' classes, a criterion of this type may be useful, e.g., if n points located on a nonlinear manifold of $\mathbb{R}^p$ are to be approximated by piecewise linear functions (smoothing and interpolation problems). Another interesting application concerns the analysis of multiplicative pseudo-random number generators where the lattice structure of consecutive numbers can be elucidated by (2.7).

The clustering criterion (2.7) may be derived from a probabilistic model as well: Consider $x_1,\ldots,x_n$ as a sample of n independent normally distributed random vectors $X_1,\ldots,X_n$ in $\mathbb{R}^p$ with the assumptions:

<u>M 1:</u> *There exists a partition $\mathcal{C} = (C_1,\ldots,C_m)$ of $\{1,\ldots,m\}$ and m unknown s-dimensional hyperplanes $H_1,\ldots,H_m \subset \mathbb{R}^p$ such that $X_k \sim N_p(\mu_k, \sigma^2 I_p)$ for all k, with unknown expectations $\mu_k = E[X_k]$ constrained by*

$$\mu_k \in H_i \quad \text{for} \quad k \in C_i\,, \quad i = 1,\ldots,m \tag{2.8}$$

*and $\sigma^2 > 0$ an unknown variance.*

The following assertion is easily proved:

Lemma 2.1: *The maximum likelihood estimate for $\mu_k$ is given by $\hat{\mu}_k = P_{H_i}(x_k)$ for all $k \in C_i$, and the optimum hyperplanes $H_i^*$ respectively the optimum partition $\mathcal{C}^*$ are determined by (2.7).*

Remark 2.1: For analysing elongated clusters FRIEDMAN/RUBIN (1967) have used the clustering criterion $\det(W(\mathcal{C})) \to \min$ which may be motivated by the model $X_k \sim N_p(z_i,\Sigma)$ for $k \in C_i$, with class-specific centers $z_i \in \mathbb{R}^p$ and an unknown, but common covariance matrix $\Sigma > 0$ (BOCK 1974, §12, §16). Evidently this model is more restriced than M 1. Another distinction resides in the fact that the source of classification and the dependence of variables are both modeled by the expectation vectors $\mu_k = E[X_k]$ in M 1 whilst in the other model they are described separately by the mean values resp. the covariance structure.

The following k-means-like algorithm is steadily decreasing the criterion (2.7) by constructing iteratively a sequence $\mathcal{C}^t, \mathcal{H}^t = (H_1^t,\ldots H_m^t)$ of partitions resp. hyperplanes $(t = 0,1,2,\ldots)$.

Algorithm A.1:

0. $t := 0$: Choose an initial m-partition $\mathcal{C}^0 = (C_1^0,\ldots,C_m^0)$.
1. Given $\mathcal{C}^t$, compute for each class $C_i^t$ a principal component hyperplane $H_i^t = \bar{x}_{C_i^t} + [v_{i1}^t,\ldots,v_{si}^t]$ using the s 'largest' eigenvectors of the scatter matrix $W(C_i^t)$.
2. Given $\mathcal{H}^t$, the partition $\mathcal{C}^{t+1}$ is the minimum-distance partition generated by the hyperplanes $H_1^t,\ldots,H_m^t$ with classes

   $$C_i^{t+1} := \{k \mid d(x_k,H_i^t) = \min_j d(x_k,H_j^t)\} \qquad i = 1,\ldots,m.$$
3. Iterate 1. and 2. until stationarity is obtained.

This algorithm improves steadily on g since 1. amounts to minimizing $g(\mathcal{C}^t,\mathcal{H})$ w.r. to $\mathcal{H}$, and 2. to minimizing $g(\mathcal{C},\mathcal{H}^t)$

w.r.t. to $\mathscr{C}$. It has been proposed by BOCK (1974, §17) and DIDAY (1979; chap. 8), and by BOCK (1979), BEZDEK et al. (1981) in a fuzzy clustering framework.

Remark 2.2: s+1 different points determine uniquely an s-dimensional hyperplane in $\mathbb{R}^p$. Therefore, a complete fit can always be attained in (2.7) if $m = [n/(s+1)] + 1$ classes are provided. This shows that the numbers m and s must be chosen with care in order to avoid trivial solutions. In particular, the class sizes $n_i$ should be large enough in order to avoid badly behaving scatter matrices $W(C_i)$.

## 2.3 PCA-based clustering with a common hyperplane

As an alternative to the probabilistic model M 1 we may consider a one-way classification model M 2 where in each class $C_i$ all means $\mu_k = E[X_k]$ are identical to some specific point $z_i$, but all of them are concentrated on the same hyperplane H:

M 2: *There is an unknown partition* $\mathscr{C} = (C_1,\ldots,C_m)$ *and an unknown hyperplane* $H = a + [v_1,\ldots,v_s]$ *such that*

$$X_k \sim N_p(z_i, \sigma^2 I_p) \qquad \text{for all} \quad k \in C_i, \ i = 1,\ldots,m$$

$$z_i \in H \qquad \text{for} \quad i = 1,\ldots,m \quad \text{(unknown)}$$

*with* $\sigma^2 > 0$ *an unknown variance.*

It is straightforward to show that maximum-likelihood estimation leads, after insertion of $a = \bar{x}$ and $z_i = P_H(\bar{x}_{C_i})$, to the optimization (clustering) criterion:

$$g(\mathscr{C},V) := \sum_{i=1}^{m} n_i \|\bar{x}_{C_i} - P_H(\bar{x}_{C_i})\|^2 + \sum_{k=1}^{m} \sum_{k \in C_i} \|x_k - \bar{x}_{C_i}\|^2 \tag{2.9}$$

$$= tr(\bar{V}'B(\mathscr{C})\bar{V}) + tr(W(\mathscr{C})) \tag{2.10}$$

$$= tr(\bar{V}'S\bar{V}) + tr(V'W(\mathscr{C})V) \to \min_{\mathscr{C},V} \tag{2.11}$$

with $V = (v_1,\ldots,v_s)$, $\bar{V} = (v_{s+1},\ldots,v_p)$ as indicated in Section 2.1. Whilst (2.10) is just the trace notation of (2.9), the last formula (2.11) derives directly from (2.9) by using the Pythagorean decomposition $\|x\|^2 = \|V'x\|^2 + \|\bar{V}'x\|^2$ for $x := x_k - \bar{x}_{C_i}$ which implies $tr(W(\mathcal{C})) = tr(V'W(\mathcal{C})V) + tr(\bar{V}'W(\mathcal{C})\bar{V})$. Remembering that $S = W(\mathcal{C}) + B(\mathcal{C})$ for all $\mathcal{C}$, we obtain (2.11).

The two expressions (2.10),(2.11) for the same criterion $g(\mathcal{C},V)$ render it possible to design an iterative relaxation algorithm for finding an (at least local) solution of (2.9). Due to its analogy with algorithms described by HUBER (1985), it may be called *projection pursuit clustering* $(t = 0,1,2,\ldots)$:

**Algorithm A.2:**

0. $t = 0$: Start with an m-partition $\mathcal{C}^0 = (C_1^0,\ldots,C_m^0)$.

1. For a given $\mathcal{C}^t$, minimize $g(\mathcal{C}^t,V)$ w.r. to $V = (v_1,\ldots,v_s)$ for obtaining an optimal hyperplane $H^t = \bar{x} + [v_1^t,\ldots,v_s^t]$. By (2.10) this is equivalent to $tr(\bar{V}'B(\mathcal{C}^t)\bar{V}) \to \min$. A look at the basic PCA problem (2.3) shows that the p-s *'smallest'* eigenvectors $v_{s+1}^t,\ldots,v_p^t$ of $B(\mathcal{C}^t)$ must be chosen for $\bar{V}$, thus $H^t$ is spanned by the s *'largest'* eigenvectors of the between scatter matrix $B(\mathcal{C}^t)$.

2. For a given hyperplane $H^t$, choose the partition $\mathcal{C}^{t+1}$ which minimizes $g(\mathcal{C},H^t)$. Since the first term in (2.11) is constant in this case and the second term is
$$tr(V^{t'}W(\mathcal{C})V^t) = \sum\sum \|V^{t'}(x_k - \bar{x}_{C_i})\|^2 \to \min_{\mathcal{C}}, \qquad (2.12)$$
this amounts to minimize the usual variance criterion (2.6) for the projected data points $y_k \in \mathbb{R}^p$, (2.4), resp. for the principal component vectors $x_k^* := V^{t'}(x_k - \bar{x}) \in \mathbb{R}^s$, $k = 1,\ldots,n$.

3. Iterate 1. and 2. until stationarity.

By construction, the values of g are steadily decreasing: $g(\mathcal{C}^t,H^t) \geq g(\mathcal{C}^t,H^{t+1}) \geq g(\mathcal{C}^{t+1},H^{t+1})$ for all t. This property holds even if we use in 2., instead of the global optimum $\mathcal{C}^{t+1}$, a local optimum obtained by the usual k-means procedure for $x_1^*,\ldots,x_n^*$ (starting with $\mathcal{C}^t$).

## 2.4 Cluster-guided PCA

If we are given a 'natural' classification $\mathcal{C} = (C_1,\ldots,C_m)$ of the data $x_1,\ldots,x_n \in \mathbb{R}^p$ we may ask which s-dimensional hyperplane $H = a + [v_1,\ldots,v_s]$ may be best suited for a visual display of this classification. Classical discriminant analysis solves this problem by maximizing the variance *between* the classes of $\mathcal{C}$:

$$k(\mathcal{C},H) := \sum_{i=1}^{m} n_i \|\overline{x}^*_{C_i} - \overline{x}^*\|^2 = tr(V'B(\mathcal{C})V) \rightarrow \max_H \qquad (2.13)$$

for the 'projected' data points $x_k^* := V'(x_k-\overline{x}) \in \mathbb{R}^s$. If, however, the 'natural' classification $\mathcal{C}$ is unknown, we have to maximize (2.13) simultaneously over H and $\mathcal{C}$: By this way, we obtain a pair $H^*,\mathcal{C}^*$ where $H^*$ is best suited to $\mathcal{C}^*$, and $\mathcal{C}^*$ is the classification which is best 'explained' by an s-dimensional linear subspace (s discrimination factors).

An iterative algorithm for solving this problem (at least approximately) has been proposed, e.g., by DIDAY (1979, chap. 9: analyse typologique discriminante). As a matter of fact, it appears that DIDAY's algorithm is identical to our projection pursuit algorithm A.2 of Section 2.3: Evidently, the step 1. of A.2 is maximizing (2.13) over H, and step 2. is maximizing (2.13) over $\mathcal{C}$ (cf. (2.12)). Thus we have seen that the same algorithm solves two different problems, one based on probabilistic model, the other on geometrical considerations.

## 2.5 PCA-clustering with common and class-specific dimensions

Class-specific hyperplanes (Section 2.2) resp. one common hyperplane only (Section 2.4) are particular cases of a more general clustering model where each class $C_i$ is characterized by a hyperplane $H_i = a_i + [u_1,\ldots,u_t,v_{i1},\ldots,v_{is}]$ spanned by s class-specific unit vectors $v_{i1},\ldots,v_{is} \in \mathbb{R}^p$ (*class-specific* factors), and t additional unit vectors $u_1,\ldots,u_t \in \mathbb{R}^p$ which are the same for all classes (*common* factors). The corresponding optimization problem leads, in analogy to (2.7), to the clustering criterion:

$$g(\mathscr{C},\mathscr{H}) := \sum_{i=1}^{m} \sum_{k\in C_i} \|(I-UU' - V_iV_i')(x_k-\bar{x}_{C_i})\|^2 \qquad (2.14)$$

$$= tr(W(\mathscr{C})) - tr(U'W(\mathscr{C})U) - \sum_{i=1}^{m} tr(V_i'W(C_i)V_i) \rightarrow \min_{\mathscr{C},\mathscr{H}}$$

where $U = (u_1,\ldots,u_s)$, $V_i = (v_{i1},\ldots,v_{is})$ etc. By minimizing in turn w.r. to $\mathscr{C}$ and $\mathscr{H}$, we can devise an iterative relaxation algorithm for solving (2.14), but computation will be much more difficult than before. Actually, for a fixed matrix U, we can prove that $v_{i1},\ldots,v_{is}$ will be the 'largest' s eigenvectors of the matrix $(I - UU')\ W(C_i)(I - UU')$, but the subsequent minimization w.r. to U necessitates a cumbersome iterative process.

## 3. CLASSIFICATION AND MULTIDIMENSIONAL SCALING

The problem raised in Sections 2.3 and 2.4 in the context of PCA may be formulated in the framework of MDS as well. After introducing some notation in Section 3.1 we will show in Section 3.2 how MDS methods can be modified in order to take account of some given classification $\mathscr{C}$, and in Section 3.3 a simultaneous clustering-and-representation algorithm will be proposed.

### 3.1 Multidimensional scaling: The scalar product version

Given an n×n dissimilarity matrix $D = (d_{k\ell})$ with $0 = d_{kk} \leq d_{k\ell} = d_{\ell k}$ for all $k,\ell$, we ask for an s-dimensional representation $x_1,\ldots,x_n \in \mathbb{R}^s$ for the n objects $1,\ldots,n$ which approximates this dissimilarity structure: The *scalar product version of MDS* looks for a representation $X = (x_1,\ldots,x_n)'$ such that the n×n scalar product matrix $\Sigma(X) := (\sigma_{k\ell}) = -1/2 \cdot E_nXX'E'$ with elements $\sigma_{k\ell} := (x_k-\bar{x})'(x_\ell-\bar{x})$ approximates as well as possible matrix

$$S := (s_{k\ell}) := -1/2 \cdot E_n D_2 E_n' \qquad (3.1)$$

with elements ('similarities'):

$$s_{k\ell} := -\frac{1}{2}(d^2_{k\ell} - \overline{d^2_{k.}} - \overline{d^2_{\ell.}} + \overline{d^2_{..}}) \qquad k,\ell \in \mathcal{O}. \qquad (3.2)$$

Here $E_n := I_n - (1/n)\mathbb{1}_{n\times n} \in \mathbb{R}^{n\times n}$ is called the centering matrix, $D_2 = (d^2_{k\ell})$ denotes the matrix of squared dissimilarities, and as usually $\overline{d^2_{k.}}, \overline{d^2_{..}}$ denote the row (total) means of the squares $d^2_{k\ell}$. Assuming (without restriction) that $\bar{x} = 0$, this problem is stated formally by:

$$\Phi(X) := \sum_{k=1}^{n} \sum_{=1}^{n} (s_{k\ell}-\sigma_{k\ell})^2 = tr([S - XX']^2) \to \min_{X\in\mathbb{R}^{n\times s}}. \qquad (3.3)$$

This formulation is motivated by the fact that in case $D = (d_{k\ell}) = (\|x_k - x_\ell\|)$ is the euclidean distance matrix of n points $x_1,\ldots,x_n \in \mathbb{R}^p$ (then D is called 'euclidean'), the indices $s_{k\ell}$, (3.2), are identical to the scalar products $\sigma_{k\ell}$, and $\Phi(X) = 0$.

The exact solution of the matrix approximation problem (3.3) is described in terms of the eigenvalues $\lambda_1 \geq \lambda_2 \geq \ldots \geq \lambda_n$ of $S = (s_{k\ell})$, the corresponding orthonormal eigenvectors $v_1,\ldots,v_n \in \mathbb{R}^n$, and the number $\rho$ of positive eigenvalues of S:

<u>*Theorem 3.1:*</u> *The optimal s-dimensional representation is given by*

$$X^* = (x^*_1,\ldots,x^*_n)' = (v_1\sqrt{\lambda_1},\ldots,v_s\sqrt{\lambda_s})' \qquad \text{if } s \leq \rho$$
$$(3.4)$$
$$X^* = (x^*_1,\ldots,x^*_n)' = (v_1\sqrt{\lambda_1},\ldots,v_\rho\sqrt{\lambda_\rho},0\ldots0) \qquad \text{if } s > \rho.$$

This has been proved by KELLER (1958), and by MATHAR (1985) for the case of orthogonally invariant deviation measures $\Phi$. Evidently $\rho$ is a 'critical' dimension since a dimension $s > \rho$ provides no better approximation than $s = \rho$.

## 3.2 MDS allowing for a given classification

Consider the case where, apart from the dissimilarity matrix $D = (d_{k\ell})$, some classification $\mathcal{C} = (C_1,\ldots,C_m)$ is known for n objects 1,...,n. (For example, $\mathcal{C}$ may have emerged from some distance based clustering algorithm.) How can we find an s-dimensional representation $x_1,\ldots,x_n \in \mathbb{R}^s$ of the objects *and* a representation $y_1,\ldots,y_m \in \mathbb{R}^s$ of the classes $C_1,\ldots,C_m$ which will optimally reproduce the dissimilarity structure between objects *and* classes?

Let us define the dissimilarity between two classes $C_i, C_j$ by

$$\overline{D^2_{C_iC_j}} := (n_in_j)^{-1} \cdot \sum_{k\in C_i} \sum_{k\in C_j} d^2_{k\ell} \tag{3.5}$$

with similar formulas for $\overline{D^2_{k,C_i}}, \overline{D^2_{C_i,\cdot}}$ etc., $D(\mathcal{C}) := (\overline{D^2_{C_iC_j}})_{m\times m}$ and the 'similarity' indices

$$\overline{S_{C_iC_j}} := (n_in_j)^{-1} \sum_{k\in C_i} \sum_{k\in C_j} s_{k\ell} = \tag{3.6}$$

$$= -\frac{1}{2}\left(\overline{D^2_{C_iC_j}} - \overline{D^2_{C_i,\cdot}} - \overline{D^2_{C_j,\cdot}} + \overline{D^2_{\cdot\cdot}}\right).$$

In matrix notation, (3.6) is expressed by

$$S(\mathcal{C}) := (\overline{S_{C_iC_j}})_{m\times m} = M'_{\mathcal{C}}\, S\, M_{\mathcal{C}} = -\frac{1}{2} E_{\mathcal{C}}\, D(\mathcal{C}) E'_{\mathcal{C}}\ . \tag{3.7}$$

Here $E_{\mathcal{C}}$ denotes the weighted centering matrix $E_{\mathcal{C}} := I_m - 1\!\!1_m q'$ with $q = (n_1,\ldots,n_m)'/n$ the vector of relative class sizes. The averaging matrix $M_{\mathcal{C}} = J_{\mathcal{C}} N^{-1}$ (with entries 0 or $1/n_i$) is defined by $N := \mathrm{diag}(n_1,\ldots,n_m)$ and the $n\times m$ matrix $J = (c_{ki}) \in \mathbb{R}^{n\times m}$ which describes the given partition $\mathcal{C}$: $c_{ki} := 1\ (0)$ iff $k \in C_i$ $(k \notin C_i)$.

That the definition (3.6) makes sense is exemplified by the case where $D = (d_{k\ell}) = (\|x_k - x_\ell\|)$ is a euclidean distance matrix: Then it appears that

$$\overline{S_{C_iC_j}} = (\bar{x}_{C_i} - \bar{x})'(\bar{x}_{C_j} - \bar{x}) \quad \text{resp.} \quad S(\mathcal{C}) = E_{\mathcal{C}} YY' E_{\mathcal{C}} \tag{3.8}$$

with $Y := (\bar{x}_{C_1},\ldots,\bar{x}_{C_m})' = M'_{\mathcal{C}} X$ the matrix of class means, and

$$\delta_{C_iC_j} := \overline{S_{C_iC_i}} + \overline{S_{C_jC_j}} - 2\cdot\overline{S_{C_iC_j}} = \|\bar{x}_{C_i} - \bar{x}_{C_j}\|^2. \tag{3.9}$$

A detailed analysis of the indices $\bar{D}, \bar{S}$, and $\delta$ is given by BOCK (1986).

We now come back to the question raised at the beginning of this Section and ask for a *simultaneous* s-dimensional representation $X = (x_1, \ldots, x_n)' \in \mathbb{R}^{n \times s}$, $Y = (y_1, \ldots, y_m)' \in \mathbb{R}^{m \times s}$ of objects and classes such that the corresponding scalar products

$$(x_k - \bar{x})'(x_\ell - \bar{x}) \qquad (x_k - \bar{x})'\,(y_i - \tilde{y}) \qquad (y_i - \tilde{y})'(y_j - \tilde{y})$$

(with $\tilde{y} := \Sigma\, n_i y_i / n$) approximate as best as possible the corresponding indices $s_{k\ell}$, $\bar{S}_{k,C_i}$, and $\bar{S}_{C_i C_j}$. (In fact they should be identical for a euclidean matrix D and $y_i := \bar{x}_{C_i}$.) Assuming without restriction that $\bar{x} = \tilde{y} = 0$, this idea may be formally stated by the optimization problem:

$$\begin{aligned} g(X,Y,\mathcal{C}) := {} & \sum_{k=1}^{n} \sum_{\ell=1}^{n} (s_{k\ell} - x_k' x_\ell)^2 \\ & + 2\alpha^2 \cdot \sum_{k=1}^{n} \sum_{j=1}^{m} \gamma_j^2 \, (\bar{S}_{k,C_j} - x_k' y_j) \\ & + \alpha^4 \cdot \sum_{i=1}^{m} \sum_{j=1}^{m} \gamma_i^2 \gamma_j^2 \, (\bar{S}_{C_i C_j} - y_i' y_j)^2 \\ = {} & \mathrm{tr}\left( \left[ \begin{pmatrix} I_n \\ \alpha \Gamma M_{\mathcal{C}}' \end{pmatrix} \cdot S \cdot (I_n \mid \alpha M_{\mathcal{C}} \Gamma) - \begin{pmatrix} X \\ \alpha \Gamma Y \end{pmatrix} \cdot (X' \mid \alpha Y' \Gamma) \right]^2 \right) \\ = {} & \mathrm{tr}\left( [\quad S^* \quad - \quad Z \quad \cdot \quad Z' \quad ]^2 \right) \\ \longrightarrow {} & \min_{X,Y} . \end{aligned} \tag{3.10}$$

Here we have introduced a weight $\alpha > 0$ for balancing between the importance of object-to-object and class-to-class relations; some weights $\gamma_j > 0$ for the classes $C_j$ of $\mathcal{C}$ with $\Gamma := \mathrm{diag}(\gamma_1, \ldots, \gamma_m)$; the matrix $Z := (x_1, \ldots, x_n, \alpha\gamma_1 y_1, \ldots, \alpha\gamma_m y_m)' = (X' \mid \alpha Y' \Gamma)' \in \mathbb{R}^{(n+m) \times s}$. In analogy to (3.3),(3.10) is a positive definite approximation problem for the matrix $S^*$

whose solution Z is described by theorem 3.1. A more detailed analysis (BOCK 1986) shows that the eigenvalues/-vectors of $S^*$ can be given in terms of the eigenvalues $\lambda_1 \geq \lambda_2 \geq \ldots \geq \lambda_n$ and of the eigenvectors $u_1, \ldots, u_n \in \mathbb{R}^n$ of the symmetric $n \times n$ matrix

$$\hat{S} := (I_n + \alpha^2 M_{\mathfrak{C}} \Gamma^2 M_{\mathfrak{C}}')^{1/2} S (I + \alpha^2 M_{\mathfrak{C}} \Gamma^2 M_{\mathfrak{C}})^{1/2} = Q^{1/2} S Q^{1/2} \quad (3.11)$$

where $Q := I_n + \alpha^2 M_{\mathfrak{C}} \Gamma^2 M_{\mathfrak{C}}' \geq I_n$ is positive definite. Denoting by $\rho$ the number of positive eigenvalues of $\hat{S}$ we have:

<u>Theorem 3.2:</u> *The solution $X^*, Y^*$ of (3.10) is given by*

$$X^* := (x_1^*, \ldots, x_n^*)' = Q^{-1/2}(u_1\sqrt{\lambda_1}, \ldots, u_s\sqrt{\lambda_s}) \qquad \text{if } s \leq \rho$$

$$(3.12)$$

$$X^* := (x_1^*, \ldots, x_n^*)' = Q^{-1/2}(u_1\sqrt{\lambda_1}, \ldots, u_\rho\sqrt{\lambda_\rho}, 0, \ldots, 0) \qquad \text{if } s > \rho$$

*and by the corresponding class means $y_i^* := \bar{x}_{C_i}$, i.e. by*

$$Y^* := (y_1^*, \ldots, y_m^*)' = M_{\mathfrak{C}}' X^*.$$

Note that the solution $X^*$ may be computed by applying the usual software packages for solving (3.3) to the matrix $\hat{S}$ and transforming by $Q^{-1/2}$.

<u>Remark 3.1:</u> The most obvious choice $\alpha = 1$, $\gamma_i \equiv 1$ assigns very different weight sums $n^2$, $2mn$, $m^2$ to the three sums in (3.10). A more balanced weighting of object-to-object and group-to-group relations is obtained for $\alpha^2 = n/m$, $\gamma_i \equiv 1$ or $\alpha = 1$, $\gamma_i = \sqrt{n_i}$ with weight sums $n^2$, $2n^2$, $n^2$. As an alternative, a sequence of increasing values $\alpha > 0$ may be used.

<u>Remark 3.2:</u> Since $\hat{S} \geq S$ we have $\lambda_k(\hat{S}) \geq \lambda_k(S)$ for all k which implies $\rho(\hat{S}) \geq \rho(S)$ for the critical dimensions of (3.10) and (3.3). Therefore cluster-based MDS (3.10) allows generally a larger dimension s than the usual MDS method (3.3) without necessitating the insertion of non-informative zero components into $X^*$.

## 3.3 MDS-based cluster analysis

Let $D = (d_{k\ell})$ be a given dissimilarity matrix for the objects $1,\ldots,n$. In the context of cluster analysis we may ask how we can find (characterize) an optimal classification $\mathcal{C} = (C_1,\ldots,C_m)$ and, *simultaneously*, an optimal representation $X = (x_1,\ldots,x_n)' \in \mathbb{R}^{n\times s}$ of the objects such that both are best suited to each other, e.g., such that $\mathcal{C}$ is most evident from a visual inspection of the points $x_1,\ldots,x_n \in \mathbb{R}^s$.

A corresponding MDS-based clustering criterion is given by

$$g(X,\mathcal{C}) := 2\beta \cdot \sum_{i=1}^{m} \sum_{k\in C_i} \|x_k - \bar{x}_{C_i}\|^2 + \mathrm{tr}([S - XX']^2) \to \min_{\mathcal{C},X} \quad (3.13)$$

where the first term evaluates the quality of the partition $\mathcal{C}$ in terms of the points $x_1,\ldots,x_n$ (i.e. *not* in terms of the given dissimilarities $d_{k\ell}$), and the second term relates to the fit between the matrix $S = (s_{k\ell})$, (3.1), and its scalar-product counterpart $XX'$ (assuming $\bar{x} = 0$). $\beta > 0$ is a weight introduced for balancing between the criteria of 'homogeneity' and 'approximation error'.

Some elementary algebra leads to the following formula for g:

$$g(X,\mathcal{C}) = \mathrm{tr}(S^2 - \Sigma_{\mathcal{C}}^2) + \mathrm{tr}([\Sigma_{\mathcal{C}} - XX']^2) \quad (3.14)$$

where the $n\times n$ matrices $R_{\mathcal{C}} := M_{\mathcal{C}} N M_{\mathcal{C}}' = J_{\mathcal{C}} N^{-1} J_{\mathcal{C}}'$ (with entries $1/n_i$ or 0) and $\Sigma_{\mathcal{C}} := S + \beta(R_{\mathcal{C}} - I_n)$ have been introduced. This is the key formula for defining an iterative relaxation algorithm A.3 of the k-means type which produces a series $\mathcal{C}^0$, $X^0$, $\mathcal{C}^1$, $X^1$, ... of partitions resp. representations:

**Algorithm A.3:**

0. $t = 0$: Start with an initial partition $\mathcal{C}^0 = (C_1^0,\ldots,C_m^0)$.

1. Given the partition $\mathcal{C}^t$, define $X^t = (x_1^t,\ldots x_n^t)'$ by minimizing $g(X,\mathcal{C}^t)$ w.r.t. to X. By (3.14) this is equivalent to the matrix approximation problem $\mathrm{tr}([\Sigma_{\mathcal{C}} - XX']^2) \to \min_X$ of

the type (3.3). Its solution is given by Theorem 3.1, where $\lambda_i = \lambda_i(\Sigma_{\mathcal{C}})$, $v_i = v_i(\Sigma_{\mathcal{C}})$, $\rho = \rho(\Sigma_{\mathcal{C}})$ now relate to $\Sigma_{\mathcal{C}}$ instead of S.

2. For a given representation $X^t$, find a new m-partition $\mathcal{C}^{t+1}$ by minimizing $g(X^t,\mathcal{C})$ w.r. to $\mathcal{C}$. Due to (3.13), $\mathcal{C}^{t+1}$ is the optimum clustering resulting from the variance criterion (2.6) for the given points $x_1^t,\ldots x_n^t \in \mathbb{R}^s$.

3. Iterate 1. and 2. for t = 0,1,2,... until stationarity.

By construction, this algorithm improves steadily on g since $g(X^t,\mathcal{C}^t) \geq g(X^t,\mathcal{C}^{t+1}) = g(X^{t+1},\mathcal{C}^{t+1})$ for all t. This holds even in the (more realistic) case where $\mathcal{C}^{t+1}$ is the sub-optimal partition obtained by the usual k-means algorithm (starting with $\mathcal{C}^t$).

Proof of (3.14): Since $\bar{x} = 0$ and $M_{\mathcal{C}}' X$ has rows $\bar{x}_{C_i}'$ we have

$$\sum\sum \|x_k - \bar{x}_{C_i}\|^2 = \sum \|x_k\|^2 - \sum n_i \|\bar{x}_{C_i}\|^2 =$$

$$= tr(XX') - tr(X' M_{\mathcal{C}} N M_{\mathcal{C}}' X) =$$

$$= tr(XX') - tr(M_{\mathcal{C}} N M_{\mathcal{C}}' XX') = tr(XX') - tr(R_{\mathcal{C}} XX').$$

Therefore, adding and subtracting $tr(\Sigma^2)$, we obtain:

$$\begin{aligned} g(X,\mathcal{C}) &= 2\beta \cdot tr(XX' - R_{\mathcal{C}} XX') + tr(S^2 - 2SXX' + (X'X)^2) \\ &= tr(S^2) - 2 \cdot tr([S + \beta R_{\mathcal{C}} - \beta I_n]XX') + tr((XX')^2) \\ &= tr(S^2 - \Sigma_{\mathcal{C}}^2) + tr([\Sigma_{\mathcal{C}} - XX']^2). \end{aligned}$$

Remark 3.3: We can prove that $R_{\mathcal{C}} \leq I_n$, and therefore $S \geq \Sigma_{\mathcal{C}}$. This shows that the critical dimension $\rho(\Sigma_{\mathcal{C}})$ for the clustering problem (3.13) is never larger than the critical dimension $\rho(S)$ of the ordinary MDS problem (3.3).

Thus the algorithm is selecting, automatically, a dimension s as small as possible.

Remark 3.4: The second (error) term in the clustering criterion (3.13) does not involve the clustering $\mathcal{C}$; in analogy to Section (3.2) it could be replaced by the criterion (3.10) if a simultaneous display $y_1,\ldots,y_m$ of the m classes is desired. The modification of the algorithm is selfevident, and Theorem 3.2 replaces Theorem 3.1 in the step 1 of algorithm A.3.

Institute of Statistics
Technical University Aachen
Wuellnerstr. 3
D-5100 Aachen/FRG

REFERENCES

Bezdek, J.C., Ch. Coray, R. Gunderson, J. Watson: Detection and characterization of cluster substructure. I. Linear structure: Fuzzy c-lines. II. Fuzzy c-varietes and convex combinations thereof. *SIAM J. Appl. Math.* 40 (1981) 339-357, 358-372.

Bock, H.H.: *Automatische Klassifikation. Theoretische und praktische Methoden zur Gruppierung und Strukturierung von Daten (Cluster-Analyse).* Vandenhoeck & Ruprecht, Göttingen 1974, 480 pp.

Bock, H.H.: Clusteranalyse mit unscharfen Partitionen. In: Bock, H.H. (ed.): *Klassifikation und Erkenntnis III: Numerische Klassifikation.* Studien zur Klassifikation SK-6. INDEKS-Verlag, Frankfurt 1979, 137-163.

Bock, H.H.: Multidimensional scaling in the framework of cluster analysis. In: Degens, P.O., H.-J. Hermes, O. Opitz (eds.): *Classification and its environment.* (Proc. 10. Annual Meeting of the Gesellschaft für Klassifikation) Studien zur Klassifikation SK-17. INDEKS-Verlag, Frankfurt 1986.

Diday, E. et al.: *Optimisation en classification automatique I,II.* Institut National de Recherche en Informatique et en Automatique (INRIA), Le Chesnay/France 1979.

Friedman, H.P., J. Rubin: On some invariant criteria for grouping data. *J. Amer. Statist. Assoc.* **62** (1967) 1159-1178.

Huber, P.J.: Projection pursuit. *Annals of Statistics* **13** (1985) 435-475.

Keller, J.B.: Factorization of matrices by least squares. *Biometrika* **49** (1962) 239-242.

Mathar, R.: The best Euclidean fit to a given distance matrix in prescribed dimensions. *Linear Algebra and Its Applications* **67** (1985) 1-6.

Okamoto, M., M. Kanazawa: Minimisation of eigenvalues of a matrix and optimality of principal components. *Ann. Math. Statist.* **39** (1968) 859-863.

Pearson, K.: On lines and planes of closest fit to systems of points in space. *Phil. Mag.* (6th series) **2** (1901) 559-572.

Rao, C.R.: The use and interpretation of principal component analysis in applied research. *Sankhyā* **A 26** (1964) 329-358.

Hamparsum Bozdogan and Donald E. Ramirez

# AN EXPERT MODEL SELECTION APPROACH TO DETERMINE THE "BEST" PATTERN STRUCTURE IN FACTOR ANALYSIS MODELS

## ABSTRACT

This paper introduces and develops an expert data-analytic model selection approach based on *Akaike's Information Criterion (AIC)* and asymptotically *Consistent Akaike's Information Criterion (CAIC)*, to choose the *number of factors*, m, and to determine the *"best" factor pattern structure* among all possible patterns under the orthogonal factor model using *Mallows' Cp Criterion*.

A subset selection procedure is carried out using a *"leaps and bounds"* algorithm to interpret the complex interrelationships between the best fitting number of factors and the original variables.

The new approach presented in this paper tries to *unify* both the *exploratory* and *confirmatory* factor analysis to find *"the best fitting simple structure for the best m-factor model"* in one expert statistical system.

Numerical examples are provided to show how to achieve flexibility in modeling and to demonstrate the efficiency of this procedure.

**Key words and phrases:**
Model Selection Criteria, AIC, CAIC and Mallows' Cp
Choosing the Number of Factors
Determining the Pattern Structure

## 1. INTRODUCTION AND PURPOSE

Factor analysis is a very useful and important multivariate statistical technique which is used to find a way of



*H. Bozdogan and A. K. Gupta (eds.), Multivariate Statistical Modeling and Data Analysis, 35–60.*

condensing the information contained in a number of original variables into a smaller set of composite factors.

In our previous work (Bozdogan and Ramirez, 1987a), we presented the well-known *information-theoretic* model selection criteria such as those of *Akaike's Information Criterion (AIC)* due to Akaike (1973), (1974), (1977), and the asymptotically *Consistent Akaike's Information Criterion (CAIC)* (see e.g., Bozdogan, 1987), for deciding how many factors to extract. In this paper we continue the use of model selection criteria in problems of factor analysis.

In *exploratory factor analysis*, the researcher has no *prior* knowledge about the variables measured nor how the factors depend upon these variables. Therefore, an exploratory analysis is carried out in two steps as explained in Lawley (1958). The first step is to decide how many factors best fit the data, and a second step consists of a *rotation* of these factors into others so that a simple but meaningful interpretation of the experimental results can be achieved.

Having decided on the *best fitting number of factors* by the minimum AIC or CAIC procedure for a given data set, we show how to carry out an *expert analytical model selection approach* for determining the *simple factor pattern structure* in factor analysis models. In this paper we will investigate how the extracted factors depend upon the original observed variables by carrying out a subset selection procedure using regression analysis. This method provides an analytical approach for interpreting the complex interrelationships of each factor with the original variables. We determine the factor pattern structure using *Mallows' (1973) Cp Criterion*. One could use the AIC or CAIC procedure in lieu of Mallows' Cp but we have used the Cp criterion because of its accessibility and the fact that it is first-order equivalent to AIC in regression models as we will show in Section 3.2.1. In Section 2, we briefly set up the factor analysis model as needed in this paper. We avoid the unnecessary generalities of the factor model taking into account the fact that the readers are familiar with the subject. In Section 3, we give the derived form of *Akaike's Informa-*

*tion Criterion (AIC)* for the *orthogonal factor model*. We estimate the factor scores by using a *ridge estimator* and address the problem of determining the pattern structure of the factor loading matrix $\Lambda$ via Mallows' Cp. In Section 4, we show numerical examples on two data sets to demonstrate these analytical techniques. The first example is a simulated example and the second is the study of Lord's (1956) speed data. In these numerical examples, our goal is to show analytically how to resolve two problems that face investigators doing factor analysis:

(i) What is the dimensionality of the factor space?

(ii) What is the pattern of free parameters in the factor model without postulating *a priori* the simple pattern structure?

In Section 5 we present our conclusions and discussion.

## 2. FACTOR ANALYSIS MODEL

Let x be a vector of the p observable variables $x_1, x_2, \ldots, x_p$, with mean vector $\mu$ which can be assumed to be zero, and covariance matrix $\Sigma$. We also assume that $\Sigma$ is of full rank p. Then we say that the *m-factor model* holds for $x_1, x_2, \ldots, x_p$ which are linear functions of the m underlying factors (where m < p), plus a residual term:

$$\begin{aligned} x_1 &= \lambda_{11}f_1+\lambda_{12}f_2+\cdots+\lambda_{1m}f_m+\varepsilon_1 \\ x_2 &= \lambda_{21}f_1+\lambda_{22}f_2+\cdots+\lambda_{2m}f_m+\varepsilon_2 \\ &\vdots \\ x_p &= \lambda_{p1}f_1+\lambda_{p2}f_2+\cdots+\lambda_{pm}f_m+\varepsilon_p, \end{aligned} \tag{2.1}$$

where the weights $\{\lambda_{im}\}$ are called the *factor loadings*, $f_1, f_2, \ldots, f_m$ represent the m *common factors*, and $\varepsilon_1, \varepsilon_2, \ldots, \varepsilon_p$ are the residual terms whose variances $\Psi_i$

represent the "uniqueness" of $x_i$. Thus, we can write the model (2.1) in matrix form as follows:

$$\underset{(p\times 1)}{x} = \underset{(p\times m)}{\Lambda}\ \underset{(m\times 1)}{f} + \underset{(p\times 1)}{\varepsilon}\ . \tag{2.2}$$

Then we say that the model in (2.2) is an *orthogonal factor model (OFM)* which imposes a *covariance structure* for x given by

(1) $$\mathrm{Cov}(x) = \Sigma = \Lambda\Lambda' + \Psi, \tag{2.3}$$

and

(2) $$\mathrm{Cov}(x,f) = \Lambda. \tag{2.4}$$

The usual assumptions for the orthogonal factor model are:

$$f \sim N_m(0, I_m),\ m \le p;$$

$$\varepsilon \sim N_p(0,\Psi),\ \text{where } \Psi = \mathrm{Diag}(\psi_1, \psi_2, \ldots, \psi_p);$$

$$f \text{ and } \varepsilon \text{ are independent; and} \tag{2.5}$$

$$\Lambda'\Psi^{-1}\Lambda \equiv \Delta,$$

where $\Delta$ is an $(m\times m)$ diagonal matrix, i.e., $\Delta = \mathrm{Diag}(\delta_1, \ldots, \delta_m)$ with $\delta_1 > \delta_2 > \cdots > \delta_m$. This is called the *matrix uniqueness condition* because of the multiplicity of choices for $\Lambda$, the factor loading matrix (see e.g., Johnson and Wichern 1982, p. 416). This constraint is used to force the iterative maximum likelihood estimates to converge to a unique solution. Also, the residual variances are usually constrained so that $\psi_i \ge \delta$ for all i, where $\delta$ is a *"small" positive value*, e.g., 0.005. If one or more of the estimated values $\hat{\psi}_i$ of $\psi_i$ take a *negative* value, or if the matrix of residual variances $\hat{\Psi}$ is not positive-definite, such a solution is clearly inadmissible and is said to be *improper*, or a *Heywood*

(1931) *case*. This may arise because of sampling errors or because the factor model is inappropriate.

In our algorithm, FACAIC (see e.g., Bozdogan and Ramirez 1987b), as we fit m = 1,2,...,M factor models, the $\Psi$ values from the previous model are used as starting values for the subsequent model to achieve much faster convergence and to obtain more stable solutions.

For more on the convergence of iterative procedures in factor analysis, we refer the reader to the work of Tumura and Sato (1981a) and (1981b).

## 3. THE MAXIMUM LIKELIHOOD FACTOR ANALYSIS

In practice, we observe a data matrix X whose information is summarized by the sample mean $\bar{x}$ and sample covariance matrix S, or sample correlation matrix R. To estimate the factor loadings, we shall assume that m is fixed in advance and that x has the p-variate normal distribution with mean vector $\mu$ and covariance matrix $\Sigma$ (positive-definite). That is,

$$x \sim N_p(\mu,\Sigma). \tag{3.1}$$

We are interested in the maximum likelihood (ML) estimates of these parameters under the *orthogonal factor model* $M_{OF}$, where

$$M_{OF}: \Sigma = \Lambda \Lambda' + \Psi. \tag{3.2}$$

The interesting problem is how to estimate $\Lambda$ and $\Psi$ (and hence $\Sigma = \Lambda \Lambda' + \Psi$) from S or R; that is, we wish to find estimates $\hat{\Lambda}$ and $\hat{\Psi}$, satisfying the *uniqueness constraint* in (2.5).

### 3.1 Identification of the Number of Factors by AIC and CAIC

For a particular data set, in practice, one does not know the "*true*" or "*actual*" number of factors m. One of the difficult and delicate tasks of factor analysis is the selection of m, the number of common factors, based on a finite set of available data. If we can adequately choose m, the number of factors, and describe the underlying process, that *best approximating m-factor* model may be useful for some future purpose.

In the literature, the classical *"goodness-of-fit"* test type procedures are often criticized on the grounds that the critical values of these test criteria are not adjusted to allow for the fact that a set of hypotheses is being tested (Everitt, 1984, p. 22; Lawley and Maxwell, 1971, p. 37). Moreover, in practice, the problem is usually not testing a particular hypothesis, but rather a *multiple decision problem* (Anderson and Rubin, 1956).

For the purpose of identifying the number of factors present for a given particular data set, we give the closed form expression of *Akaike's Information Criterion (AIC)* for the *orthogonal factor model*.

The AIC statistic in general is defined by

$$\mathrm{AIC}(k) = -2\log L[\hat{\Theta}(k)]+2k, \tag{3.3}$$

where $L[\Theta(k)]$ is the likelihood function of the observations, $\hat{\Theta}(k)$ is the maximum likelihood estimate of the parameter vector $\Theta$ (in this case $\Theta = (\mu,\Lambda,\Psi)$), and k is the number of independent parameters estimated within the model. Here "log" is the natural logarithm.

For the orthogonal factor model $M_{OF}$ in (3.2), the first component of AIC is given by

$$-2\log L(\hat{\mu},\hat{\Lambda},\hat{\Psi}) = np\log(2\pi)+n\log|\hat{\Lambda}\hat{\Lambda}'+\hat{\Psi}|+np. \tag{3.4}$$

Counting the number of parameters, under the orthogonal factor model, the factor loading matrix $\Lambda$ has mp parameters to be estimated and the specific factor matrix, $\Psi$, has p parameters to be estimated, for a total of (mp+p) parameters. However, the uniqueness condition $\Lambda'\Psi^{-1}\Lambda \equiv \Delta$ involves $m(m+1)/2$ distinct parameters and $\Delta$ is constrained to be diagonal. Hence, this condition imposes $m(m+1)/2-m = m(m-1)/2$ additional constraints on the problem. Effectively, under the orthogonal factor model (3.2), $\Sigma$ has

$$s = (mp+p) - \frac{1}{2} m(m-1) \tag{3.5}$$

*free parameters.*

Thus, AIC under this model is given by

$$\begin{aligned} AIC(m) = {} & np \log(2\pi) + n \log|\hat{\Lambda}\hat{\Lambda}'+\hat{\Psi}|+np \\ & + 2[(mp+p) - \frac{1}{2} m(m-1)], \end{aligned} \tag{3.6}$$

where

- $n$ = the total sample size,
- $p$ = the number of original variables,
- $\hat{\Lambda}$ = maximum likelihood factor loading matrix,
- $\hat{\Psi}$ = maximum likelihood specific factor matrix,
- $m$ = number of factors, $m = 1,2,...,M$.

For a fixed number of p original variables, we fit the number of factors m not exceeding the largest integer M such that

$$m \leq M \leq \frac{1}{2}[2p+1-(8p+1)^{1/2}]. \tag{3.7}$$

We call this bound *big-factor*. It is the largest number of factors we can allow and still have the number of free parameters (3.5) in the factor model be less than the number of parameters in the covariance matrix.

Derivation of the asymptotically *Consistent Akaike's Information Criterion (CAIC)* given in Bozdogan (1987) follows similarly, except we replace the penalty component in AIC in (3.6) by s log(n), that is, the number of free parameters times the natural logarithm of the sample size n.

Since the maximum likelihood factor model is scale invariant, we use the correlation matrix R in our estimation and in extracting the factors.

After the correct number of factors is chosen, the correlation structure is rotated according to the *varimax rotation procedure* of Kaiser (1958). For more on model selection approach to the factor model problem, and on the empirical performances of AIC and CAIC in choosing the number of factors under several different factor models, we refer the reader to Bozdogan and Ramirez (1987a).

## 3.2 How the Factors Depend upon the Original Variables

As we mentioned in the introduction, the interpretation of factor analysis results is very difficult and sometimes almost impossible. For example, if we *postulate* factor pattern structure in advance based on *prior* knowledge, we can generate a large number of such factor patterns. Indeed, the total number of confirmatory models possible is $2^{pm}-1$. We prefer to determine the "*best*" factor pattern structure empirically from the data rather than postulating a specific factor pattern. We have found that the factor patterns are quite useful in interpreting the factor loadings. As McDonald and Burr (1967) point out, the regression estimators of the factor scores are useful when the investigator wishes the estimated factor scores to have high correlations with the "*true*" factor scores (see also Mulaik, 1972). In fact, the method of nonlinear factor analysis described by McDonald (1962, 1967) requires the knowledge of factor scores.

It is well known that factor scores cannot be determined precisely, but can only be "*estimated*," since the common factors do not fully account for the total variance of the variables. If we adopt this notion and invoke some "*minimum variance*" or "*least squares*" principle, then reasonable estimates of factor scores may be obtained. There are, of course, many different ways one can estimate the factor scores. The problem is to select which estimation method one should use. For more on these issues, we refer the reader to Harris (1967) or McDonald and Burr (1967).

For our purpose, in this paper we compute the factor scores and use them:

(1) to check the underlying assumptions of the factor model in (2.5),

(2) to help identify the zeros in the factor loading matrix, and
(3) to explain the existence of improper solutions when the factor model is overfitted.

3.2.1 The Estimation of Factor Scores and Subset Selection by Mallows' Cp. So far we studied and gave the closed form analytical expression of AIC on how to choose m, the number of factors in the factor model. We now reverse the problem and look at how the factors depend upon the observed variables and compute the factor scores F. It is well known that F cannot be estimated in the usual fashion since it is a random vector and the m+n unknown random variables F and $\varepsilon$ exceed the number of observations n.

As discussed in Mardia *et al.* (1979), it would be theoretically more correct and attractive to estimate the *factor scores*, *factor loadings*, and *specific variances* all simultaneously from the given data, using the maximum likelihood method. However, there are parameter values for which the likelihood function becomes infinite. To see this, let m = 1 and suppose $\Lambda = \lambda$ (p×1) and $\mu = 0$ are known. Let F (n×1) denote the factor scores. Then the likelihood of the data X given the F's can be written as

$$L(X|F) = |2\pi\Psi|^{-1/2}\exp\left[-\frac{1}{2}\sum_{i=1}^{p}(X_i-\lambda_i F)'(X_i-\lambda_i F)/\Psi_{ii}\right]. \qquad (3.8)$$

Suppose $\lambda_1 \neq 0$. It can be shown that for any values of $\Psi_{22},\cdots,\Psi_{pp}$ if $F = \lambda_1^{-1}X_1$ and $\Psi_{11} \rightarrow 0$, then the likelihood becomes infinite. Hence, the MLE's do not exist in this case. Therefore, we treat F as fixed and try to estimate it from the same data used to estimate the parameters $\mu$, $\Lambda$, and $\Psi$.

There are several methods for estimating the factor scores, and there has been a long controversy as to which procedure is best. Following Seber (1984, p. 220), we use a ridge method of estimation of the factor scores which will have a smaller *mean square error* than the commonly used Bartlett's (1937, 1938) generalized least squares estimate.

If the linear estimate $\hat{f}$ of f is chosen to minimize the mean square error $E[\|\hat{f}-f\|^2]$, then $\hat{f}$ is a *ridge estimator* for the conditional model

$$x|f \sim N(\Lambda f, \Psi), \tag{3.9}$$

and is given by

$$\underset{(m\times 1)}{\hat{f}} = \underset{(m\times p)}{\Lambda'} \underset{(p\times p)}{R^{-1}} \underset{(p\times 1)}{x} = \underset{(m\times m)}{(I+\Delta)^{-1}} \underset{(m\times p)}{\Lambda'} \underset{(p\times p)}{\Psi^{-1}} \underset{(p\times 1)}{x} . \tag{3.10}$$

The matrix of estimated factor scores is thus

$$\underset{(n\times 1)\ \cdots\ (n\times 1)}{[\hat{F}_1|\hat{F}_2|\cdots|\hat{F}_m]}$$

$$= \underset{(n\times m)}{\hat{F}} = \underset{(n\times p)}{X} \underset{(p\times p)}{\hat{R}^{-1}} \underset{(p\times m)}{\hat{\Lambda}} = \underset{(n\times p)}{X} \underset{(p\times m)}{\hat{B}} . \tag{3.11}$$

So,

$$\hat{F}_r = X\hat{R}^{-1}\lambda_r, \tag{3.12}$$

with $\lambda_r$ the r-th column of the factor loading matrix $\hat{\Lambda}$, $(1 \le r \le m)$.

Hence $\hat{f}$ is conditionally biased as

$$E[\hat{f}|f] \neq f, \tag{3.13}$$

and, as with ridge estimators, the elements of $\hat{f}$ will have a smaller mean square error than those of *Bartlett's factor scores*.

We now use Mallows' Cp criterion

$$Cp = 1/\sigma^2\ RSS(p)-n+2p \tag{3.14}$$

an estimate of mean-square error, divided by $\sigma^2$, in subset selection to identify which of the original variables best load on each of the common factors chosen to be optimal. This yields the *pattern structure* for the factor loadings matrix $\Lambda$. We view (3.11) as a regression model with estimator of B given by

$$\hat{B} = \hat{R}^{-1}\Lambda = (\Lambda\Lambda' + \Psi)^{-1}\Lambda. \quad (3.15)$$

For us, the work-horse subroutine which computes the best subset from all the possible regressions is the IMSL subroutine RLEAP which uses a "*leaps and bounds*" algorithm to find the best subset regressions for a full regression model. Here "*best*" is defined according to the minimum of Mallows' Cp criterion. This is a computationally efficient algorithm which appears to be a *minimum of arithmetic*, that is, less than six floating-point operations per regression, which is based on a "*leaps and bounds*" technique for finding the best subsets without examining all possible subsets. For a detailed explanation of this algorithm, we refer the reader to Furnival and Wilson (1974).

In equation (3.14) $\sigma^2$ is estimated by $\hat{\sigma}^2 = RSS(p)/n$, which is usually taken to be the estimate of the mean square error when all the variables are in the model, n is the number of observations, and RSS(p) is the residual sum of squares.

In the case of linear regression models, there is a relationship between AIC and Cp statistic. It is easy to see this relationship if we write AIC as

$$AIC(p) = n \log(\hat{\sigma}^2)+2p \quad (3.16)$$

for the regression model. Rewriting (3.16) in the equivalent form, we can have

$$AIC(p) = n \log(\hat{\sigma}^2)+2p-n \log(\sigma^2) \quad (3.17)$$

$$= n \log[\hat{\sigma}^2/\sigma^2]+2p.$$

Therefore, we have firmed up the following folklore result.

Theorem 3.2.1. AIC(p) and Cp are *"first order equivalent"* in the case of linear regression models.

Proof.
$$AIC(p) = n\ \log[\hat{\sigma}^2/\sigma^2]+2p \tag{3.18}$$

$$= n\ \log[\frac{(\hat{\sigma}^2-\sigma^2)+\sigma^2}{\sigma^2}]+2p$$

$$= n\ \log[1+[(\hat{\sigma}^2-\sigma^2)/\sigma^2]]+2p$$

$$\doteq n\ [(\hat{\sigma}^2-\sigma^2)/\sigma^2]+2p$$

$$= n[\hat{\sigma}^2/\sigma^2]-n+2p = Cp. \quad \Box$$

It is remarkable that Cp is applied only in the case of linear regression models whereas AIC is used in more general modeling situations such as the one presented in this paper.

## 4. NUMERICAL EXAMPLES

In this section we give two numerical examples to demonstrate the expert model selection approach for determining the *"best" factor pattern structure* in *orthogonal factor models*.

All the computations are carried out by using our FACAIC and FSCORE algorithms in double precision on one of the PRIME 750 computers of the University of Virginia. For the FORTRAN 77 source code for these algorithms, we refer the reader to Bozdogan and Ramirez (1987b), (1987c), copies of which can be obtained from the authors upon request.

TABLE I
The AIC's, CAIC's and the P-Values for $\chi^2$ for the Simulated Data Set

| Number of Factors m | Number of Parameters s | AIC(m) | CAIC(m) | P-Values for $\chi^2$ |
|---|---|---|---|---|
| 1 | 24 | 3021.3 | 3083.8 | 0.0000 |
| 2 | 35 | 2827.6 | 2918.8 | 0.0000 |
| 3* | 45 | 2737.7* | 2854.9** | 0.4802 |
| 4# | 54 | 2738.1 | 2878.7 | 0.8750 |
| 5 | 62 | 2744.6 | 2906.1 | 0.9597 |
| 6 | 69 | 2752.7 | 2932.4 | 0.9868 |
| 7 | 75 | 2763.0 | 2958.4 | 0.8596 |
| 8 | 80 | 2772.3 | 2980.7 | *** |
| 0 | 78 | 2768.2 | 2971.4 | *** |

Note: $n = 100$ and $p = 12$.
The $m = 0$ case is the unrestricted model.
*Minimum AIC at $\hat{m} = 3$ factors.
**Minimum CAIC at $\hat{m} = 3$ factors.
***P-Values of $\chi^2$ cannot be computed due to nonpositive d.f.
#Improper solutions occur when fitting $m = 4$-factor model.

### 4.1 Example 1: A Simulated Example

This example is important in that it clearly demonstrates the power and versatility of our approach in recovering the "*true*" *factor pattern structure* we constructed.

We simulated n = 100 observations from a p = 12 variate multinormal distribution with mean vector $\boldsymbol{\mu} = 0$ and covariance matrix $\boldsymbol{\Sigma} = \Lambda\,\Lambda' + \Psi$ with

$$\underset{(12\times3)}{\Lambda} = \begin{bmatrix} .9 & 0 & 0 \\ .9 & 0 & 0 \\ .9 & 0 & 0 \\ .9 & 0 & 0 \\ 0 & .8 & 0 \\ 0 & .8 & 0 \\ 0 & .8 & 0 \\ 0 & .8 & 0 \\ 0 & 0 & .7 \\ 0 & 0 & .7 \\ 0 & 0 & .7 \\ 0 & 0 & .7 \end{bmatrix}, \qquad (4.1)$$

and

$$\underset{(12\times12)}{\Psi} = \mathrm{Diag}(.19,.19,.19,.19,.36,.36,$$

$$.36,.36,.51,.51,.51,.51). \qquad (4.2)$$

So, the covariance matrix has the following (4×4) block-diagonal form $\Sigma_{ii}$ repeated three times:

$$\Sigma_{ii} = \begin{bmatrix} 1.00 & & & \\ .81 & 1.00 & & \\ .81 & .81 & 1.00 & \\ .81 & .81 & .81 & 1.00 \end{bmatrix} \qquad (4.3)$$

The simulation was done using the IMSL subroutine GGNSM with initial seed = 9849006825.

Our results for choosing the best fitting m-factor model using AIC and CAIC are given in Table I.

Looking at Table I, we see that the minimum AIC and CAIC both occur at m = 3-factor model, indicating that three factors are needed to explain the simulated data adequately. Note that in this example $m^* = 3$ was the actual number of true factors.

After choosing $m = 3$-factor model as the best fitting factor model, the correlation structure is rotated according to the *varimax rotation procedure*. The factor loadings matrix, $\Lambda$, using the varimax rotation is computed to be:

$$\underset{(12\times 3)}{\hat{\Lambda}} = \begin{bmatrix} -0.895 & 0.089 & 0.062 \\ -0.912 & 0.151 & -0.002 \\ -0.911 & 0.064 & 0.077 \\ -0.907 & 0.099 & 0.093 \\ 0.030 & -0.849 & 0.022 \\ 0.135 & -0.783 & 0.027 \\ 0.078 & -0.836 & 0.002 \\ 0.113 & -0.773 & 0.015 \\ -0.034 & 0.080 & -0.681 \\ 0.086 & 0.040 & -0.765 \\ 0.093 & 0.008 & -0.640 \\ 0.031 & -0.067 & -0.679 \end{bmatrix} \qquad (4.4)$$

and

$$\hat{\Psi} = \text{Diag}(.187, .145, .159, .159, .278, .367, .295, .389, .529, .405, .582, .533). \qquad (4.5)$$

As Jöreskog (1978) notes:

> The factor analyst usually only interprets the large loadings in a factor matrix $\Lambda$ after rotation to simple structure. However, the most difficult question which has plagued factor analysts for several decades is, "How large should a factor loading be to be considered significant?" or, "How small should a factor loading be to be ignored?"

We now determine how the extracted factors depend upon the observed variables by the subset selection on the best subset from all the possible regressions by using the "leaps and bounds" technique in our FSCORE algorithm. The results of the computed "best" subset factor score weights for each of the three factors are given by

$$\underset{(12\times 3)}{\hat{B}} = \begin{bmatrix} -0.224 & 0 & 0 \\ -0.295 & 0 & 0 \\ -0.271 & -0.073 & 0 \\ -0.260 & 0 & 0 \\ -0.087 & -0.338 & 0 \\ 0 & -0.229 & 0 \\ 0 & -0.312 & 0 \\ 0 & -0.215 & 0 \\ 0 & 0 & -0.267 \\ 0 & 0 & -0.379 \\ 0 & 0 & -0.220 \\ -0.041 & 0 & -0.264 \end{bmatrix} \tag{4.6}$$

The zeros in the factor score weights in (4.6) help to place the zeros in the factor loadings matrix. Thus, the *"best" factor pattern structure* is now specified and the investigator can use this pattern as an initial pattern for a confirmatory factor model.

Using (4.6) we estimate factor scores from $\hat{F} = X\hat{B}$. We then do some exploratory data analysis to check the factor model assumptions given in (2.5). In particular, we compute the correlation $\mathrm{Cor}(\hat{E},\hat{F})$ between the estimated errors and the estimated factor scores, and the correlation $\mathrm{Cor}(X,\hat{F})$ between the original variables and the estimated factor scores.

TABLE II
Examining the Improper Solutions for the Simulated Data Set

| | $h_i^2 = 1-\hat{\Psi}_i$ = Communality Estimates | | | | | | | | | | | |
|---|---|---|---|---|---|---|---|---|---|---|---|---|
| m/p | 1 | 2 | 3 | 4 | 5 | 6 | 7 | 8 | 9 | 10 | 11 | 12 |
| 1 | 813 | 844 | 836 | 840 | 021 | 056 | 036 | 046 | 000 | 017 | 018 | 008 |
| 2 | 813 | 844 | 843 | 842 | 722 | 632 | 702 | 611 | 012 | 026 | 022 | 008 |
| 3 | 813 | 854 | 841 | 841 | 722 | 633 | 705 | 610 | 471 | 595 | 418 | 467 |
| 4# | 811 | 881 | 849 | 838 | 1000 | 588 | 858 | 558 | 473 | 594 | 458 | 457 |

Note: #Improper solutions occur when fitting $m = 4$-factor model at variable 5.

Best fitting model at $\hat{m} = 3$-factor model by both AIC and CAIC.

Maximum possible communality is normalized to 1000.

$$\mathrm{Cor}(\hat{E},\hat{F}) = \begin{bmatrix} -.177 & -.034 & -.028 \\ -.060 & -.069 & -.057 \\ .008 & -.097 & -.057 \\ -.216 & -.034 & -.022 \\ -.080 & -.103 & -.025 \\ .001 & -.244 & -.014 \\ -.016 & -.227 & -.033 \\ -.016 & -.159 & -.012 \\ -.035 & -.024 & -.214 \\ -.056 & -.010 & -.394 \\ -.042 & -.019 & -.140 \\ -.077 & -.005 & -.222 \end{bmatrix} \qquad \mathrm{Cor}(X,\hat{F}) = \begin{bmatrix} -.921 & .108 & .106 \\ -.939 & .171 & .049 \\ -.932 & .072 & .121 \\ -.933 & .117 & .133 \\ .036 & -.896 & .041 \\ .158 & -.837 & .033 \\ .107 & -.889 & .013 \\ .136 & -.820 & .027 \\ -.012 & .093 & -.759 \\ .106 & .056 & -.871 \\ .120 & .015 & -.709 \\ .036 & -.047 & -.761 \end{bmatrix} \tag{4.7}$$

From (4.7) one can visually observe that the elements of $\mathrm{Cor}(\hat{E},\hat{F})$ are small and that $\mathrm{Cor}(X,\hat{F})$ agrees with $\hat{\Lambda}$ given in (4.4). This verifies the assumptions in (2.5).

In Table II, we report the communality estimates as the number of factors increase.

From Table II, we observe that the solutions become improper once the model has been overfitted. Using the estimated factor scores with m = 4, the correlation between the estimated errors and the factor scores are seen to be different from zero with values over 0.6, (e.g., $\mathrm{Cor}(\hat{E}_5,\hat{F}_3) = -.69$).

A simple explanation of this improper solution can now be given. After the factor model has been correctly fitted, that is, after we choose the best fitting factor model, the additional factors are not true factors but rather are the residual errors from a lower dimensional model. For example, the correlation between the error for $X_5$ in the model with m = 3 and the fourth factor in the model with m = 4 is -.93. Thus the error for $X_5$ with m = 3 has become the fourth factor with m = 4 and the overfitted model becomes improper.

TABLE III

The AIC's, CAIC's and the P-Values for $\chi^2$ for Lord's Speed Factor Data

| Number of Factors m | Number of Parameters s | AIC(m) | CAIC(m) | P-Values for $\chi^2$ |
|---|---|---|---|---|
| 1 | 30 | 24608 | 24742 | 0.0000 |
| 2 | 44 | 21637 | 21834 | 0.0000 |
| 3 | 57 | 20846 | 21102 | 0.0000 |
| 4 | 69 | 20730* | 21039** | 0.0177 |
| 5# | 80 | 20716 | 21074 | 0.5449 |
| 6# | 90 | 20722 | 21124 | 0.7470 |
| 0 | 120 | 20757 | 21294 | *** |

Note: n = 649 and p = 15.

The m = 0 case is the unrestricted model.

*Minimum AIC for proper $\hat{m}$ = 4-factor model.

**Minimum CAIC for proper $\hat{m}$ = 4-factor model.

***P-Values of $\chi^2$ cannot be computed due to nonpositive d.f.

#Improper solutions occur when fitting m = 5,6-factor models.

## 4.2 Example 2: Lord's Data on Speed Factors

To further demonstrate how the computing of the factor score weight matrix $\hat{B}$ is useful in exploratory data analysis in an orthogonal factor model, we analyze a subset of p = 15 variables from 39 reported by Lord (1956). We will use his test scores on *vocabulary*, namely, $X_3$, $X_4$, $X_6$, $X_7$, $X_8$; test scores on *visualizing intersections of geometric objects*: $X_{11}$, $X_{12}$, $X_{14}$, $X_{15}$, $X_{16}$; and test scores on *arithmetic reasoning*: $X_{19}$, $X_{20}$, $X_{22}$, $X_{23}$, $X_{24}$.

The respondents were n = 649 students at the U.S. Naval Academy in Annapolis.

Table III shows the AIC and CAIC values up to m = 4-factor model, since improper solutions occur when m = 5.

Looking at Table IV, we see that both the AIC and CAIC have picked the m = 4-factor model as the best fitting model. For this model, the estimated factor loading matrix $\hat{\Lambda}$, and the estimated factor score weight matrix $\hat{B}$ are given in Table IV.

From Table IV, it is easy to identify factor 1 as "*vocabulary*," factor 2 as "*intersections*," and factor 3 as "*arithmetic reasoning*." We are not surprised by the small negative weights in the $\hat{B}$ matrix. For example, the arithmetic reasoning variables are acting as *suppressor variables* for the regression equation that predicts the vocabulary factor (see e.g., Lord and Novick (1968) for a discussion of the role of suppressor variables in regression models).

We now turn our attention to the fourth factor and the corresponding fourth column of the factor score coefficient or weight matrix $\hat{B}$ in Table IV. This factor has the undesirable property of being loaded on nearly all of the original X-variables. But note that there are two *clusters*: the *first cluster* with the *positive signs* ($X_4$, $X_5$, $X_9$, $X_{10}$, $X_{13}$, $X_{15}$); and, the *second cluster* with

TABLE IV
Estimated Factor Loading and Factor Score Weight Matrices

| | | $\hat{\lambda}$ | | |
|---|---|---|---|---|
| p/m | 1 | 2 | 3 | 4 |
| 1 | **.746** | .034 | .173 | -.181 |
| 2 | **.768** | .091 | .172 | -.266 |
| 3 | **.798** | .098 | .218 | .077 |
| 4 | **.904** | .021 | .207 | .108 |
| 5 | **.863** | .017 | .236 | .200 |
| 6 | .075 | **.769** | .212 | -.159 |
| 7 | .023 | **.818** | .165 | -.073 |
| 8 | .082 | **.848** | .183 | .018 |
| 9 | .051 | **.890** | .162 | .065 |
| 10 | .027 | **.877** | .234 | .103 |
| 11 | .183 | .225 | **.639** | -.173 |
| 12 | .170 | .205 | **.660** | -.167 |
| 13 | .199 | .170 | **.764** | .134 |
| 14 | .202 | .224 | **.716** | .046 |
| 15 | .229 | .135 | **.730** | .087 |

| | | $\hat{B}$ | | |
|---|---|---|---|---|
| p/m | 1 | 2 | 3 | 4 |
| 1 | **.135** | 0 | 0 | -.288 |
| 2 | **.184** | 0 | -.053 | -.506 |
| 3 | **.150** | 0 | 0 | 0 |
| 4 | **.412** | 0 | -.117 | .245 |
| 5 | **.284** | -.022 | -.047 | .491 |
| 6 | 0 | **.139** | 0 | -.254 |
| 7 | 0 | **.170** | -.043 | -.136 |
| 8 | 0 | **.214** | -.061 | 0 |
| 9 | 0 | **.315** | -.109 | .158 |
| 10 | 0 | **.302** | 0 | .285 |
| 11 | -.028 | -.024 | **.189** | -.195 |
| 12 | -.034 | -.030 | **.206** | -.193 |
| 13 | -.086 | -.071 | **.345** | .184 |
| 14 | -.058 | -.043 | **.259** | 0 |
| 15 | -.058 | -.062 | **.276** | .093 |

the *negative signs* ($X_1$, $X_2$, $X_6$, $X_7$, $X_{11}$, $X_{12}$). This condition warns the researcher that the factor 4 is an *artifactor*. The careful investigator will return to the data and discover that all of the variables in the first cluster were collected in an "*untimed*" or no speed environment while all of the variables in the second cluster were collected in a "*timed*" or highly speeded environment.

Thus the factor score coefficient matrix $\hat{B}$ has led the investigator to a missing but important covariate. Jöreskog (1979) has analyzed this data set allowing the low-speed and high-speed items to have their own covariance structure.

## 5. CONCLUSIONS AND DISCUSSION

In this paper we have introduced a new expert data-analytic model selection approach; (1) to determine how many factors to fit to a given multivariate data set, (2) to find the extent to which each variable depends on each common factor by carrying out a subset selection procedure to interpret the complex interrelationships of each factor with the original variables, and (3) to determine a "*best*" factor pattern structure in factor analysis models among all possible confirmatory models objectively.

With our new approach, the investigator can now perform both exploratory and confirmatory factor analysis by the aid of the analytical model selection procedures: AIC, CAIC, and Mallows' Cp.

Even though the approach presented in this paper tends to "*automate*" model fitting and evaluation in one "*expert statistical system*," the main objective considered was to remove the subjectivity involved in the choice of the number of factors, and to design a pattern matrix for the factor loadings to obtain a "*best fitting simple structure*" as a means for finding interpretable factors. We note here that our procedure produces a "*best fitting simple pattern structure*" marginally. It is not claimed to be the "*best*" jointly. The latter requires the modification index of Jöreskog and Sörbom (1981), and development of a new branch and bound algorithm as a guide for searching jointly. However, such an approach from a numerical analysis point of view is presently infeasible and very costly as is the case in using LISREL. Our technique is very computer efficient, less costly, and is a good place to start.

Furthermore, our procedure does not need the exact number and location of the zeros, and it does not look at the names of the original variables. It is unique up to the *varimax rotation* procedure of Kaiser (1958) which we used in our numerical examples. Further research is presently being carried out to show the effect of the other transformations in determining the "best" fitting simple factor pattern structure. The result of this research will be reported elsewhere.

We emphasize the fact that by introducing AIC, CAIC, and Mallows' Cp type model selection procedures, it is not necessary to carry out the classical hypothesis testing which is inappropriate within the context of factor model problems; it is not necessary to subjectively choose the number of factors based on an arbitrarily chosen level of significance; and it is unnecessary to subjectively construct a factor pattern structure. Thus, using our new data-analytic procedure, the investigator can achieve the desired flexibility in model fitting, and he can look at model construction and evaluation problems more comprehensively without inherent ambiguity.

In concluding, it is clear that any psychological interpretation of the factors is largely subjective. Our analytical procedure attempts to eliminate this subjectivity and human bias.

## ACKNOWLEDGEMENTS

This research was partially supported by NIH Biomedical Research Support Grant (BRSG) No. 5-24867 at the University of Virginia. We thank Professors Stan Sclove, Arjun K. Gupta, and Donald Jensen for their valuable comments.

Department of Mathematics
Math/Astronomy Building
University of Virginia
Charlottesville, VA 22903

## REFERENCES


Akaike, H. (1973). 'Information Theory and an Extension of the Maximum Likelihood Principle.' In B. N. Petrov and F. Csaki (eds.). Second International Symposium on Information Theory. Academiai Kiado: Budapest, 267-281.

Akaike, H. (1974). 'A New Look at the Statistical Model Identification.' IEEE Transactions on Automatic Control, AC-19, 716-723.

Akaike, H. (1977). 'On Entropy Maximization Principle.' In P. R. Krishnaiah (ed.), Proc. Symposium on Applications of Statistics, North-Holland, Amsterdam, 27-47.

Anderson, T. W. and Rubin, H. (1956). 'Statistical Inference in Factor Analysis.' In Proc. of the Third Berkeley Symposium on Mathematical Statistics and Probability, 5, 111-150.

Bartlett, M. S. (1937). 'The Statistical Conception of Mental Factors.' The British Journal of Psychology, **28**, 97-104.

Bartlett, M. S. (1938). 'Methods of Estimating Mental Factors.' Nature, 141, 609-610.

Bozdogan, H. (1987). 'Model Selection and Akaike's Information Criterion (AIC): the General Theory and its Analytical Extensions.' Psychometrika (to appear).

Bozdogan, H. and Ramirez, D. E. (1987a). 'Model Selection Approach to the Factor Model Problem, Parameter Parsimony, and Choosing the Number of Factors.' Annals of the Institute of Statistical Mathematics, Paper B (to appear).

Bozdogan, H. and Ramirez, D. E. (1987b). 'FACAIC: Model Selection Algorithm for the Orthogonal Factor Model Using AIC and CAIC.' Psychometrika: Computational Psychometrics Section (to appear).

Bozdogan, H. and Ramirez, D. E. (1987c). 'Factor Pattern Structure for Orthogonal Factor Analysis Model Using FSCORE.' Technical Paper No. 17 in Statistics, Department of Mathematics, University of Virginia, Charlottesville, VA 22903.

Everitt, B. S. (1984). *An Introduction to Latent Variable Models*. London: Chapman and Hall.

Furnival, G. M. and Wilson, R. W. (1974). 'Regressions by Leaps and Bounds.' *Technometrics*, **16**, 499-511.

Harris, C. W. (1967). 'On Factors and Factor Scores.' *Psychometrika*, 32, 363-379.

Haywood, H. B. (1931). 'On Finite Sequences of Real Numbers.' *Proc. of the Royal Society, Series A*, **134**, 486-501.

Johnson, R. A. and Wichern, D. W. (1982). *Applied Multi variate Statistical Analysis*. Englewood Cliffs: Prentice-Hall.

Jöreskog, K. G. (1978). 'Structural Analysis of Covariance and Correlation Matrices.' *Psychometrika*, **4**, 443-477.

Jöreskog, K. G. (1979). *Advances in Factor Analysis and Structural Equation Models*. Cambridge: Abt Books.

Jöreskog, K. G. and Sörbom, D. (1981). *LISREL V, Analysis of Linear Relationships by Maximum Likelihood and Least Squares Methods*. Uppsala, Sweden: University of Uppsala.

Kaiser, H. F. (1958). 'The Varimax Criterion for Analytic Rotation in Factor Analysis.' *Psychometrika*, **23**, 187-200.

Lawley, D. N. (1958). 'Estimation in Factor Analysis Under Various Initial Assumptions.' *British Journal of Statistical Psychology*, **11**, 1-12.

Lawley, D. N. and Maxwell, A. E. (1971). Factor Analysis as a Statistical Method. New York: American Elsevier.

Lord, F. M. (1956). 'A Study of Speed Factors in Tests and Academic Grades.' Psychometrika, 21, 31-50.

Lord, F. M. and Novick, M. R. (1968). Statistical Theories of Mental Test Scores. Reading: Addison-Wesley.

Mallows, C. L. (1973). 'Some Comments on Cp.' Technometrics, 15, 661-675.

Mardia, K. V., Kent, J. T., and Bibby, J. M. (1979). Multivariate Analysis. New York: Academic Press.

Mulaik, S. A. (1972). The Foundations of Factor Analysis. New York: McGraw Hill.

McDonald, R. P. (1962). 'A General Approach to Nonlinear Factor Analysis.' Psychometrika, 27, 397-415.

McDonald, R. P. (1967). 'Numerical Methods for Polynomial Models in Nonlinear Factor Analysis.' Psychometrika, 32, 77-112.

McDonald, R. P. and Burr, E. J. (1967). 'A Comparison of Four Methods of Constructing Factor Scores.' Psychometrika, 32, 381-401.

Seber, G. A. F. (1984). Multivariate Observations. New York: John Wiley and Sons.

Tumura, T. and Sato, M. (1981a). 'On the Convergence of Iterative Procedures in Factor Analysis.' TRU Mathematics 17-1, 159-168.

Tumura, T. and Sato, M. (1981b). 'On the Convergence of Iterative Procedures in Factor Analysis (II).' TRU Mathematics 17-2, 255-271.

Vernon M. Chinchilli

# BLUS RESIDUALS IN MULTIVARIATE LINEAR MODELS


## ABSTRACT

The examination of residuals is always an important aspect of fitting data to statistical models in terms of identifying influential observations and detecting violations of assumptions. The latter use is difficult to perform for the ordinary residuals in the univariate linear model because these residuals are not independent. This led researchers to consider alternative sets of residuals, such as the best linear, unbiased, scalar-type variance (BLUS) residuals. In this article the definition of BLUS residuals is extended to multivariate models. The extension is relatively straightforward for the multivariate analysis of variance (MANOVA) model, but not for the generalized multivariate analysis of variance (GMANOVA) model and the mixed MANOVA-GMANOVA model. For each of the GMANOVA and mixed MANOVA-GMANOVA models two sets of BLUS residuals arise naturally, namely, a "between" set and a "within" set.


Key words and phrases: MANOVA; GMANOVA; residual analysis.

## 1. INTRODUCTION

When analyzing data sets within the framework of univariate and multivariate linear models, statisticians rely upon the set of residuals for diagnostic purposes. Residuals can detect outliers and influential observations, the level of multicollinearity among the independent variables, model inadequacy, and violation of assumptions. This report concentrates on two uses of residuals in some generalized multivariate linear models: (a) constructing formal tests of hypotheses about model assumptions, such as underlying normality and homogeneity of variance; (b) conducting maximum likelihood estimation in a two-stage process called



*H. Bozdogan and A. K. Gupta (eds.), Multivariate Statistical Modeling and Data Analysis, 61–75.*

residual analysis.

The ordinary set of residuals has one disadvantage in that its elements are not independent. Theil (1965) introduced the concept of the best linear, unbiased, scalar-type variance (BLUS) residuals for the univariate linear model, called the analysis of variance (ANOVA) model here. This is reviewed along with the multivariate linear multivariate analysis of variance (MANOVA) model in the next two sections. The set of BLUS residuals is ideal for testing model assumptions and conducting an appropriate residual analysis.

The observations and the ordinary residuals enjoy a one-to-one correspondence, so that the ordinary residuals are useful in identifying outliers and influential observations. Unfortunately, the set of BLUS residuals is useless in this respect, because the one-to-one correspondence between observations and residuals is obliterated.

The construction of BLUS residuals and residual analysis are considered for the generalized multivariate analysis of variance (GMANOVA) model in Section 4 and the mixed MANOVA-GMANOVA model in Section 5. Finally, a numerical example is provided in Section 6.

## 2. THE ANOVA MODEL

### 2.1. BLUS Residuals

Consider the ANOVA model

$$\mathbf{Y} = \mathbf{X}\boldsymbol{\beta} + \boldsymbol{\varepsilon}, \tag{1}$$

where $\mathbf{Y}$ is the n-vector of observations of the dependent variable, $\mathbf{X}$ is the known $n \times r$ design matrix of full rank r ($\leq n-1$), $\boldsymbol{\beta}$ is the unknown r-vector of parameters, and $\boldsymbol{\varepsilon}$ is the unobservable n-vector of random errors, whose rows are independent and identically distributed as $N(0,\sigma^2)$ random variables.

The maximum likelihood (ML) estimators of $\boldsymbol{\beta}$ and $\sigma^2$ are found as

$$\hat{\boldsymbol{\beta}} = (\mathbf{X}'\mathbf{X})^{-1}\mathbf{X}'\mathbf{Y} \text{ and } \hat{\sigma}^2 = n^{-1}\mathbf{Y}'\mathbf{D}\mathbf{Y}, \tag{2}$$

and the n-vector of residuals is defined as

$$\mathbf{R} = \mathbf{Y}-\mathbf{X}\hat{\boldsymbol{\beta}} = \mathbf{DY}, \tag{3}$$

where

$$\mathbf{D} = \mathbf{I}_n-\mathbf{X}(\mathbf{X}'\mathbf{X})^{-1}\mathbf{X}' \tag{4}$$

and $\mathbf{I}_n$ is the $n \times n$ identity matrix. Clearly, the residuals are normally distributed with

$$E(\mathbf{R}) = \mathbf{0} \text{ and } Var(\mathbf{R}) = \sigma^2\mathbf{D}. \tag{5}$$

Although the residuals in (3) are correlated, Anscombe and Tukey (1963) stated that this fact has minimal impact in the evaluation of residual plots. However, for small values of n-r, the lack of independence could invalidate a test of normality applied to the residuals.

For this reason, Theil (1965, 1968) and Grossman and Styan (1972) considered an alternative set of residuals. Let $\mathbf{L}$ be an $n \times n-r$ matrix such that

$$M(\mathbf{L}) = M(\mathbf{D}) \text{ and } \mathbf{L}'\mathbf{L} = \mathbf{I}_{n-r}, \tag{6}$$

where $M(\cdot)$ denotes the space spanned by the argument's column vectors.

Lemma. For any $n \times n-r$ matrix $\mathbf{L}$ which satisfies the conditions in (6), $\mathbf{LL}' = \mathbf{I}_n-\mathbf{X}(\mathbf{X}'\mathbf{X})^{-1}\mathbf{X}' = \mathbf{D}$.

Proof: Because of (6), each column vector of $\mathbf{L}$ can be expressed as a linear combination of the column vectors of $\mathbf{D}$, i.e., there exists an $n \times n-r$ matrix $\mathbf{Q}$ of full rank n-r such that $\mathbf{L} = \mathbf{DQ}$. Also, there exists an $n \times r$ matrix $\mathbf{L}^*$ such that $M(\mathbf{L}^*) = M(\mathbf{X})$ and $\mathbf{L}^{*\prime}\mathbf{L}^* = \mathbf{I}_r$. Therefore, there exists a nonsingular $r \times r$ matrix $\mathbf{Q}^*$ such that $\mathbf{L}^* = \mathbf{XQ}^*$. But $\mathbf{L}^{*\prime}\mathbf{L}^* = \mathbf{I}_r$ implies that $\mathbf{Q}^*\mathbf{Q}^{*\prime} = (\mathbf{X}'\mathbf{X})^{-1}$. Next, note that $\mathbf{L}'\mathbf{L}^* = \mathbf{Q}'\mathbf{DXQ}^* = \mathbf{0}$, so that this and the previous results reveal that $[\mathbf{L}\ \mathbf{L}^*]$ is an orthogonal matrix. So $\mathbf{LL}' + \mathbf{L}^*\mathbf{L}^{*\prime} = \mathbf{I}_n$ implies that $\mathbf{LL}' = \mathbf{I}_n-\mathbf{XQ}^*\mathbf{Q}^{*\prime}\mathbf{X}' = \mathbf{D}$. This concludes the proof.

Now let $\mathbf{A}$ be any fixed $n \times m$ matrix. Das Gupta (1977) showed that $E(\mathbf{A}'\mathbf{Y}) = \mathbf{0}$ iff $M(\mathbf{A}) \subset M(\mathbf{L})$. Thus, $E(\mathbf{A}'\mathbf{Y}) = \mathbf{0}$ and $Var(\mathbf{A}'\mathbf{Y}) = \sigma^2\mathbf{I}_m$ iff $M(\mathbf{A}) \subset M(\mathbf{L})$ and $\mathbf{A}'\mathbf{A} = \mathbf{I}_m$. As seen from (5) and (6), the residuals defined in (3) do not satisfy these conditions.

Now we define the class

$$A = \{\mathbf{A}: \mathbf{A} \text{ is } n \times n-r,\ M(\mathbf{A}) = M(\mathbf{L}),\ \mathbf{A}'\mathbf{A} = \mathbf{I}_{n-r}\} \tag{7}$$

and within this class we want to find the matrix **A** which yields that **A'Y** is the "best" approximation to **H'ε**, where H is n × n-r and $\mathbf{H'H} = \mathbf{I}_{n-r}$. If **H** ε *A*, then setting **A** = H yields the "best" approximation to **H'ε** because $\mathbf{H'\varepsilon} = \mathbf{H'(Y - X\beta)} = \mathbf{H'Y}$. If $\mathbf{H} \notin A$, then the matrix **A** is taken as that member of *A* which minimizes trace{*Var*(**A**'Y-H'**ε**)}. In either case, $Var(\mathbf{A'Y}) = \sigma^2 \mathbf{I}_{n-r}$ because **A** ε *A*, so the above authors labeled these the "best" of the linear unbiased estimators with "scalar" variance (BLUS).

For checking assumptions with residuals from the ANOVA model, it is reasonable to take **H** = **L**, where L is defined in (6). Therefore, the following BLUS residuals are proposed for these purposes:

$$R^* = \mathbf{L'Y} \text{ and } R^* \sim N_{n-r}(\mathbf{0}, \sigma^2 \mathbf{I}_{n-r}). \qquad (8)$$

In order to check the normality assumption for the ANOVA model, any of the available tests of normality (see Mardia 1980 for a review) could be applied to the residuals **R*** in (8). Such tests applied to the residuals **R** in (3) are appropriate only in the asymptotic sense because $Corr(\mathbf{R}) \to 0$ as $n \to \infty$. Likewise, tests of homogeneity of variance can be conducted by partitioning the sample of n experimental units into m groups, constructing a set of BLUS residuals for each of the m groups, and then applying the appropriate test to the BLUS residuals. See Muirhead (1982, Chapter 11) for a discussion of Bartlett's modified likelihood ratio (LR) test or Brown and Forsythe (1974) for some robust tests.

## 2.2. Residual Analysis

Suppose that the design matrix **X** and the parameter vector **β** are partitioned as $\mathbf{X} = [\mathbf{X}_1 \ \mathbf{X}_2]$ and $\boldsymbol{\beta} = [\boldsymbol{\beta}_1' \ \boldsymbol{\beta}_2']'$, where $\mathbf{X}_1$ is $n \times r_1$ and $\mathbf{X}_2$ is $n \times r_2$, $r = r_1 + r_2$. Then the ANOVA model in (1) can be expressed as

$$\mathbf{Y} = \mathbf{X}_1\boldsymbol{\beta}_1 + \mathbf{X}_2\boldsymbol{\beta}_2 + \boldsymbol{\varepsilon}. \qquad (9)$$

The ML estimator of **β** in terms of the partitioned linear model in (9) is

$$\begin{bmatrix} \hat{\boldsymbol{\beta}}_1 \\ \hat{\boldsymbol{\beta}}_2 \end{bmatrix} = \begin{bmatrix} \mathbf{X}_1'\mathbf{X}_1 & \mathbf{X}_1'\mathbf{X}_2 \\ \mathbf{X}_2'\mathbf{X}_1 & \mathbf{X}_2'\mathbf{X}_2 \end{bmatrix}^{-1} \begin{bmatrix} \mathbf{X}_1'\mathbf{Y} \\ \mathbf{X}_2'\mathbf{Y} \end{bmatrix} = \begin{bmatrix} (\mathbf{X}_1'\mathbf{D}_2\mathbf{X}_1)^{-1}\mathbf{X}_1'\mathbf{D}_2\mathbf{Y} \\ (\mathbf{X}_2'\mathbf{D}_1\mathbf{X}_2)^{-1}\mathbf{X}_2'\mathbf{D}_1\mathbf{Y} \end{bmatrix} =$$

$$\begin{bmatrix} (\mathbf{X}_1'\mathbf{X}_1)^{-1}\mathbf{X}_1'\{\mathbf{I}_n - \mathbf{X}_2(\mathbf{X}_2'\mathbf{D}_1\mathbf{X}_2)^{-1}\mathbf{X}_2'\mathbf{D}_1\}\mathbf{Y} \\ (\mathbf{X}_2'\mathbf{X}_2)^{-1}\mathbf{X}_2'\{\mathbf{I}_n - \mathbf{X}_1(\mathbf{X}_1'\mathbf{D}_2\mathbf{X}_1)^{-1}\mathbf{X}_1'\mathbf{D}_2\}\mathbf{Y} \end{bmatrix}, \qquad (10)$$

where $\mathbf{D}_1$ and $\mathbf{D}_2$ are analogous to the expression for $\mathbf{D}$ in (4) with $\mathbf{X}_1$ and $\mathbf{X}_2$ replacing $\mathbf{X}$, respectively.

Residual analysis refers to the two-stage analysis in which the vector of residuals $\mathbf{R}_1 = \mathbf{D}_1\mathbf{Y}$ is (a) constructed from fitting the model $\mathbf{Y} = \mathbf{X}_1\boldsymbol{\beta}_1 + \boldsymbol{\varepsilon}$ and then (b) used as the dependent variable in fitting the univariate linear model

$$\mathbf{R}_1 = \mathbf{X}_2\boldsymbol{\beta}_2 + \boldsymbol{\varepsilon}_1. \tag{11}$$

A few researchers have considered such an approach when $\boldsymbol{\beta}_1$ represents covariates or blocks whose effects are to be removed prior to examining the important effects represented by $\boldsymbol{\beta}_2$. In this case the ML estimator is the generalized least squares (GLS) estimator $\hat{\boldsymbol{\beta}}$ in (10) because $Var(\mathbf{R}_1) = \sigma^2\mathbf{D}_1$. Unfortunately, Freund, et al. (1961) and Zyskind (1963) discussed residual analysis in terms of a least squares (LS) approach, which obviously is not efficient. The LS estimator of $\boldsymbol{\beta}_2$ in residual analysis is biased unless $\mathbf{X}_2'\mathbf{X}_1 = \mathbf{0}$, because

$$E\{(\mathbf{X}_2'\mathbf{X}_2)^{-1}\mathbf{X}_2'\mathbf{R}_1\} = (\mathbf{X}_2'\mathbf{X}_2)^{-1}\mathbf{X}_2'\mathbf{D}_1\mathbf{X}_2\boldsymbol{\beta}_2. \tag{12}$$

A modification of residual analysis employing BLUS residuals eliminates the confusion because independence yields the equivalence of the ML and LS estimators. In the first stage, the BLUS residuals $\mathbf{R}_1^*$ are constructed as $\mathbf{L}_1'\mathbf{Y}$ from fitting the linear model $\mathbf{Y} = \mathbf{X}_1\boldsymbol{\beta}_1 + \boldsymbol{\varepsilon}$, where $\mathbf{L}_1$ is defined as in (6) with $\mathbf{X}_1$ replacing $\mathbf{X}$ and $r_1$ replacing $r$. The second stage consists of fitting the linear model

$$\mathbf{R}_1^* = \mathbf{L}_1'\mathbf{X}_2\boldsymbol{\beta}_2 + \boldsymbol{\varepsilon}_1^*. \tag{13}$$

From the lemma in Section 2.1, it is seen that $\mathbf{L}_1\mathbf{L}_1' = \mathbf{D}_1$, so that the ML and LS estimator of $\boldsymbol{\beta}_2$ with the model in (13) is $\boldsymbol{\beta}_2$ in (10).

## 3. THE MANOVA MODEL

The MANOVA model is given by

$$\mathbf{Y} = \mathbf{X}\boldsymbol{\beta} + \boldsymbol{\varepsilon}, \tag{14}$$

where $\mathbf{Y}$ is the $n \times p$ response matrix, $\mathbf{X}$ is the known $n \times r$ design matrix of full rank $r$ ($\leq n-p$), $\boldsymbol{\beta}$ is the unknown $r \times p$

parameter matrix, and $\boldsymbol{\varepsilon}$ is the n × p matrix of unobservable random errors whose rows are independent and identically distributed as $N_p(0, \boldsymbol{\Sigma})$, where $\boldsymbol{\Sigma}$ is a p × p positive definite matrix. An equivalent form for describing the variance structure of $\boldsymbol{\varepsilon}$ is

$$Var\{\text{Vec}(\boldsymbol{\varepsilon})\} = \boldsymbol{\Sigma}\otimes\mathbf{I}_n, \tag{15}$$

where Vec(·) is the function which concatenates the column vectors of the matrix argument into one large column vector.

The ML estimators of $\boldsymbol{\beta}$ and $\boldsymbol{\Sigma}$, $\hat{\boldsymbol{\beta}}$ and $\hat{\boldsymbol{\Sigma}}$, and the n × p residual matrix, $\mathbf{R}$, are

$$\hat{\boldsymbol{\beta}} = (\mathbf{X}'\mathbf{X})^{-1}\mathbf{X}'\mathbf{Y}, \ \hat{\boldsymbol{\Sigma}} = n^{-1}\mathbf{Y}'\mathbf{D}\mathbf{Y}, \text{ and } \mathbf{R} = \mathbf{D}\mathbf{Y}, \tag{16}$$

where $\mathbf{D}$ is defined in (4). Also,

$$E(\mathbf{R}) = \mathbf{0} \quad \text{and} \quad Var\{\text{Vec}(\mathbf{R})\} = \boldsymbol{\Sigma}\otimes\mathbf{D}. \tag{17}$$

Because (16)-(17) are analogous to the univariate versions in (2)-(5), extending the results of the previous section to the multivariate case is straightforward.

The n × n-r matrix $\mathbf{L}$ is as defined in (6) and the BLUS residuals are constructed as

$$\mathbf{R}^* = \mathbf{L}'\mathbf{Y} \quad \text{and} \quad \text{Vec}(\mathbf{R}^*) \sim N(0, \boldsymbol{\Sigma}\otimes\mathbf{I}_{n-r}). \tag{18}$$

If a test of p-variate normality is desirable, then it should be applied to the BLUS residuals $\mathbf{R}^*$ and not $\mathbf{R}$, as the n rows of $\mathbf{R}$ are not independent. See Mardia (1980) for a review of such tests, which include tests based on multivariate versions of skewness and kurtosis, Kolmogorov-Smirnov, Cramer-von Mises, and Shapiro-Wilk. Tests of homogeneity of variance matrices, such as Bartlett's modified LR test (Muirhead, 1982), can be applied to m sets of BLUS residuals in a manner similar to that discussed in Section 2.1.

A two-stage residual analysis can be performed when the MANOVA model described in (14) is partitioned in the same manner as the model in (9). Just as in the univariate case, a biased estimator of $\hat{\boldsymbol{\beta}}_2$ results if it is performed on the set of ordinary residuals $\mathbf{R}_1 = \mathbf{D}_1\mathbf{Y}$ with LS estimation unless $\mathbf{X}_2'\mathbf{X}_1 = 0$. Instead, if the BLUS residuals $\mathbf{R}_1^* = \mathbf{L}_1'\mathbf{Y}$ are used, and the model $\mathbf{R}_1^* = \mathbf{L}_1'\mathbf{X}_2\boldsymbol{\beta}_2 + \boldsymbol{\varepsilon}_1^*$ is fit, then the ML and LS estimators of $\boldsymbol{\beta}_2$ coincide.

## 4. THE GMANOVA MODEL

The generalized multivariate analysis of variance (GMANOVA) model is given as

$$\mathbf{Y} = \mathbf{X}\boldsymbol{\beta}\mathbf{P} + \boldsymbol{\varepsilon}, \tag{19}$$

where **Y**, **X**, and **ε** are the same as for the MANOVA model described in (14), **β** is an $r \times q$ parameter matrix, and **P** is a known $q \times p$ matrix of full rank $q$ ($\leq p$). **X** is called the between-units design matrix and **P** is called the within-units design matrix. This model is especially useful for modeling repeated measurements data as a function of time or dose. The multivariate linear model is a special case of the GMANOVA model when **P** is a square matrix.

The current form of the GMANOVA model was introduced by Potthoff and Roy (1964). Rao (1965, 1967), Khatri (1966), and Grizzle and Allen (1969) were the early researchers who developed statistical inference under the GMANOVA model. The ML estimators of **β** and **Σ** are

$$\hat{\boldsymbol{\beta}} = (\mathbf{X}'\mathbf{X})^{-1}\mathbf{X}'\mathbf{Y}S^{-1}\mathbf{P}'(\mathbf{P}S^{-1}\mathbf{P}')^{-1} \text{ and}$$

$$\hat{\boldsymbol{\Sigma}} = n^{-1}\{S + \mathbf{W}'\mathbf{Y}'\mathbf{X}(\mathbf{X}'\mathbf{X})^{-1}\mathbf{X}'\mathbf{Y}\mathbf{W}\}, \tag{20}$$

where

$$S = \mathbf{Y}'\mathbf{D}\mathbf{Y},\ \mathbf{W} = \mathbf{I}_p - S^{-1}\mathbf{P}'(\mathbf{P}S^{-1}\mathbf{P}')^{-1}\mathbf{P}, \tag{21}$$

and **D** is defined in (4). Although $\hat{\boldsymbol{\beta}}$ is not normally distributed unless **P** is a square matrix, it is unbiased and

$$Var\{\mathrm{Vec}(\hat{\boldsymbol{\beta}})\} = \frac{n-r-1}{n-r-p+q-1}\,(\mathbf{P}\boldsymbol{\Sigma}^{-1}\mathbf{P}')^{-1} \otimes (\mathbf{X}'\mathbf{X})^{-1}. \tag{22}$$

The usual set of residuals could be constructed as $\mathbf{R} = \mathbf{Y} - \mathbf{X}\hat{\boldsymbol{\beta}}\mathbf{P}$. However, this proves to be fruitless, because **R** is not normally distributed and its variance matrix is not of the "scalar-type" $\boldsymbol{\Gamma} \otimes \mathbf{I}_n$. Rather, the conditional inferential approach to the GMANOVA model, as discussed by the above authors, results in two natural sets of residuals, namely, the "between" set and the "within" set.

Let $\mathbf{Q} = [\mathbf{Q}_1\ \mathbf{Q}_2]$ be a full rank $p \times p$ matrix such that $\mathbf{Q}_1$ is $p \times q$, $\mathbf{Q}_2$ is $p \times p{-}q$, $\mathbf{P}\mathbf{Q}_1 = \mathbf{I}_q$, and $\mathbf{P}\mathbf{Q}_2 = \mathbf{0}$. The GMANOVA model in (19) is transformed to

$$[Z_1 \ Z_2] = [YQ_1 \ YQ_2] \sim N\left([X\beta \ 0], \begin{bmatrix}\Gamma_{11} & \Gamma_{12}\\ \Gamma_{21} & \Gamma_{22}\end{bmatrix} \otimes I_n\right) \quad (23)$$

such that $\Gamma = Q'\Sigma Q$. Notice that only $Z_1$ involves the parameter matrix $\beta$, but that $Z_2$ can not be ignored because $Z_1$ and $Z_2$ are not independent. This suggests examining the joint distribution of $[Z_1 \ Z_2]$ in terms of the conditional distribution of $Z_1$ given $Z_2$ and the marginal distribution of $Z_2$. Therefore, according to well-known results on conditional distributions within a normally distributed random matrix (Muirhead 1982, Chapter 1),

$$[Z_1 | Z_2 \ \ Z_2] \sim N\left([X\beta + Z_2\gamma \ \ 0], \begin{bmatrix}\Gamma_{11|2} & 0\\ 0 & \Gamma_{22}\end{bmatrix} \otimes I_n\right), \quad (24)$$

where

$$\gamma = \Gamma_{22}^{-1}\Gamma_{21} \quad \text{and} \quad \Gamma_{11|2} = \Gamma_{11} - \Gamma_{12}\Gamma_{22}^{-1}\Gamma_{21} \quad (25)$$

are $(p-q) \times q$ and $q \times q$, respectively. The conditional design matrix for $Z_1$ given $Z_2$ is $[X \ Z_2]$, which is an $n \times r+(p-q)$ matrix of full rank in probability, and the ML estimator of $[\beta' \ \gamma']'$ is found in a manner analogous to the derivation of (10).

Next, let

$$\begin{aligned} E &= I_n - [X \ \ Z_2]\begin{bmatrix}X'X & X'Z_2\\ Z_2'X & Z_2'Z_2\end{bmatrix}^{-1}\begin{bmatrix}X'\\ Z_2'\end{bmatrix} \\ &= D - DZ_2(Z_2'DZ_2)^{-1}Z_2'D. \end{aligned} \quad (26)$$

Notice that E is an idempotent matrix of rank $n-r-(p-q)$ in probability. Also, E is invariant to the choice of the transformation matrix Q. Then (24) and (26) suggest the matrix of residuals $[R \ \ Z_2] = [EZ_1 \ Z_2]$, with

$$[R | Z_2 \ \ Z_2] \sim N\left([0 \ \ 0], \begin{bmatrix}\Gamma_{11|2} \otimes E & 0\\ 0 & \Gamma_{22} \otimes I_n\end{bmatrix}\right). \quad (27)$$

However, the residuals $EZ_1$ conditional on $Z_2$ are not of the "scalar-type", and unconditionally $EZ_1$ is not normally distributed as E is a function of $Z_2$.

Following the set of arguments in Section 2.1, BLUS residuals for the GMANOVA are constructed as

$$R^* = M'Z_1 \quad \text{and} \quad Z_2, \quad (28)$$

where M is an $n \times n-r-(p-q)$ matrix such that $M(M) = M(E)$ and

$\mathbf{M'M} = \mathbf{I}_{n-r-p+q}$. Then

$$\mathbf{R}^* | \mathbf{Z}_2 \sim N(\mathbf{0}, \mathbf{\Gamma}_{11|2} \otimes \mathbf{I}_{n-r-p+q}), \tag{29}$$

but this conditional distribution is free of $\mathbf{Z}_2$ so it is also the unconditional distribution. In summary, $\mathbf{R}^*$ in (28) and $\mathbf{Z}_2$ in (23) are independent with

$$\mathbf{R}^* \sim N(\mathbf{0}, \mathbf{\Gamma}_{11|2} \otimes \mathbf{I}_{n-r-p+q}) \text{ and } \mathbf{Z}_2 \sim N(\mathbf{0}, \mathbf{\Gamma}_{22} \otimes \mathbf{I}_n). \tag{30}$$

$\mathbf{R}^*$ can be called the set of "between" residuals and $\mathbf{Z}_2$ the set of "within" residuals. Note that $\mathbf{R}^*$ is invariant to the choice of the transformation matrix $\mathbf{Q}$, but $\mathbf{Z}_2$ is not. Although $\mathbf{R}^*$ in (28) has not been considered previously in the literature, $\mathbf{Z}_2$ has been employed for testing the adequacy of the within portion of the GMANOVA model.

For instance, Grizzle and Allen (1969) proposed testing $H_0$: $\{E(\mathbf{Z}_2) = \mathbf{0}\}$ versus $H_1$: $\{E(\mathbf{Z}_2) \neq \mathbf{0}\}$ under the transformed model in (23). Rejection of $H_0$ indicates that the matrix $\mathbf{P}$ in (19) is inadequate for describing the within portion of the design. An invariant test of $H_0$ is based on the eigenvalues of $\mathbf{S}_H\mathbf{S}_E^{-1}$, where

$$\mathbf{S}_H = \mathbf{Z}_2'\mathbf{X}(\mathbf{X'X})^{-1}\mathbf{X'Z}_2 \quad \text{and} \quad \mathbf{S}_E = \mathbf{Z}_2'\mathbf{DZ}_2 \tag{31}$$

are independent Wishart matrices with r and n-r degrees of freedom, respectively. The eigenvalues of $\mathbf{S}_H\mathbf{S}_E^{-1}$ are invariant to the choice of the transformation matrix $\mathbf{Q}$. (For a discussion of the multivariate tests based on these eigenvalues, see Chapter 10 of Muirhead, 1982.)

Now suppose that the between-units design matrix $\mathbf{X}$ and the parameter matrix $\boldsymbol{\beta}$ are partitioned as $\mathbf{X} = [\mathbf{X}_1 \ \mathbf{X}_2]$ and $\boldsymbol{\beta} = [\boldsymbol{\beta}_1' \ \boldsymbol{\beta}_2']'$, where $\mathbf{X}_1$ is $n \times r_1$ and $\mathbf{X}_2$ is $n \times r_2$, $r = r_1 + r_2$. Thus, the GMANOVA model in (19) is expressed as

$$\mathbf{Y} = \mathbf{X}_1\boldsymbol{\beta}_1\mathbf{P} + \mathbf{X}_2\boldsymbol{\beta}_2\mathbf{P} + \boldsymbol{\varepsilon}, \tag{32}$$

and the ML estimator of $\boldsymbol{\beta}$ is

$$\begin{bmatrix} \hat{\boldsymbol{\beta}}_1 \\ \hat{\boldsymbol{\beta}}_2 \end{bmatrix} = \begin{bmatrix} (\mathbf{X}_1'\mathbf{D}_2\mathbf{X}_1)^{-1}\mathbf{X}_1'\mathbf{D}_2 \\ (\mathbf{X}_2'\mathbf{D}_1\mathbf{X}_2)^{-1}\mathbf{X}_2'\mathbf{D}_1 \end{bmatrix} \mathbf{YS}^{-1}\mathbf{P}'(\mathbf{PS}^{-1}\mathbf{P}')^{-1}, \tag{33}$$

where $\mathbf{D}_1$ and $\mathbf{D}_2$ are analogous to $\mathbf{D}$ in (4), with $\mathbf{X}_1$ and $\mathbf{X}_2$ replacing $\mathbf{X}$, respectively. The latter result follows from

(10) and (20). A two-stage residual analysis can be performed for the GMANOVA model in terms of the transformed model described in (23) and (24). Let $\mathbf{R}_1 = \mathbf{E}_1\mathbf{Z}_1$ be the residuals from fitting the conditional model

$$\mathbf{Z}_1 = \mathbf{X}_1\boldsymbol{\beta}_1 + \mathbf{Z}_2\boldsymbol{\gamma} + \boldsymbol{\varepsilon}, \text{ given } \mathbf{Z}_2, \tag{34}$$

in the first stage, where

$$\mathbf{E}_1 = \mathbf{D}_1 - \mathbf{D}_1\mathbf{Z}_2(\mathbf{Z}_2'\mathbf{D}_1\mathbf{Z}_2)^{-1}\mathbf{Z}_2\mathbf{D}_1 \tag{35}$$

is an $n \times n$ idempotent matrix. Because the rows of $\mathbf{R}_1$ are not independent, we construct the BLUS residuals $\mathbf{R}_1^* = \mathbf{M}_1'\mathbf{Y}$, analogous to $\mathbf{R}^*$ in (28), where $\mathbf{M}_1$ is an $n \times n{-}r_1{-}(p{-}q)$ such that $\mathcal{M}(\mathbf{M}_1) = \mathcal{M}(\mathbf{E}_1)$ and $\mathbf{M}_1'\mathbf{M}_1 = \mathbf{I}_{n-r_1-p+q}$. Then the second stage consists of fitting the conditional model

$$\mathbf{R}_1^* = \mathbf{M}_1'\mathbf{X}_2\boldsymbol{\beta}_2 + \boldsymbol{\varepsilon}_1^*, \text{ given } \mathbf{Z}_2. \tag{36}$$

Then the LS and ML of $\boldsymbol{\beta}_2$ from the two-stage approach in (34)-(36) coincides with the ML estimator of $\boldsymbol{\beta}_2$ in (33) for the model in (32). This follows from the lemma in Section 2.1 and the result in (10).

## 5. THE MIXED MANOVA-GMANOVA MODEL

The mixed MANOVA-GMANOVA model, introduced by Chinchilli and Elswick (1985), is an alternative to the partitioned model in (32). In that model the two sets of effects contain the within-units design matrix $\mathbf{P}$, but in some applications it may be desirable to use a within-units design for only one set of effects. An example of this is presented in the next section. The mixed MANOVA-GMANOVA model is given as

$$\mathbf{Y} = \mathbf{X}_1\boldsymbol{\beta}_1 + \mathbf{X}_2\boldsymbol{\beta}_2\mathbf{P} + \boldsymbol{\varepsilon}, \tag{37}$$

where $\mathbf{Y}$, $\mathbf{X}_1$, $\mathbf{X}_2$, $\boldsymbol{\beta}_2$, $\mathbf{P}$, and $\boldsymbol{\varepsilon}$ are defined in the same manner as for the model in (32), and $\boldsymbol{\beta}_1$ is an $r_1 \times p$ parameter matrix. As before, $\mathbf{X} = [\mathbf{X}_1\ \mathbf{X}_2]$ denotes the concatenated design matrix.

The ML estimators of $\boldsymbol{\beta}_1$, $\boldsymbol{\beta}_2$, and $\boldsymbol{\Sigma}$ are

$$\begin{aligned}\hat{\boldsymbol{\beta}}_1 = {} & (X^{11}X_1'+X^{12}X_2')YS^{-1}P'(PS^{-1}P')^{-1}P + \\ & + (X_1'X_1)^{-1}X_1'YW,\end{aligned} \tag{38a}$$

$$\hat{\beta}_2 = (X^{21}X_1'+X^{22}X_2')YS^{-1}P'(PS^{-1}P')^{-1}, \tag{38b}$$

and

$$\hat{\Sigma} = n^{-1}\{S+W'Y'(D_2-D)YW\}, \tag{38c}$$

where $S$ and $W$ are defined in (21), $D$ in (4), $D_2$ analogous to $D$ in (4) with $X_2$ replacing $X$, and

$$\begin{bmatrix} X_1'X_1 & X_1'X_2 \\ X_2'X_1 & X_2'X_2 \end{bmatrix}^{-1} = \begin{bmatrix} X^{11} & X^{12} \\ X^{21} & X^{22} \end{bmatrix}. \tag{39}$$

Two goodness-of-fit tests are of interest within the framework of the mixed MANOVA-GMANOVA model. The first one tests $H_0$: {GMANOVA} vs. $H_1$: {mixed MANOVA-GMANOVA}, and the other tests $H_0$: {mixed MANOVA-GMANOVA} vs. $H_1$: {MANOVA}. The details are given in Chinchilli and Elswick (1985).

The construction of BLUS residuals for the mixed model takes the same conditional approach as that for the GMANOVA model in the previous section. Analogous to (23) and (24),

$$[Z_1|Z_2 \quad Z_2] \sim N\Big([X_1\beta_{11}+X_2\beta_2+Z_2\gamma \quad X_1\beta_{12}], \begin{bmatrix} \Gamma_{11\cdot 2} & 0 \\ 0 & \Gamma_{22} \end{bmatrix} \otimes I_n\Big), \tag{40}$$

where $[\beta_{11} \quad \beta_{12}] = \beta_1[Q_1 \quad Q_2]$, $\Gamma = Q'\Sigma Q$, and $\gamma$ and $\Gamma_{11\cdot 2}$ are defined in (25). The conditional design matrix for $Z_1$ given $Z_2$ is $[X \; Z_2] = [X_1 \; X_2 \; Z_2]$ and the design matrix for $Z_2$ is $X_2$ so that the ML estimators of $[\beta_{11}' \; \beta_2' \; \gamma']'$ and $\beta_{12}$ are found in the usual manner.

Now let $E$ be as defined in (26) with $X = [X_1 \; X_2]$ and let $D_1$ be defined analogously to $D$ in (4) with $X_1$ replacing $X$. Then the ordinary residuals for this model are $[R \quad D_1Z_2]$ $= [EZ_1 \quad D_1Z_2]$, where

$$[EZ_1|Z_2 \quad D_1Z_2] \sim N\Big([0 \; 0], \begin{bmatrix} \Gamma_{11\cdot 2}\otimes E & 0 \\ 0 & \Gamma_{22}\otimes D_1 \end{bmatrix}\Big). \tag{41}$$

However, the members of $EZ_1$ are not independent, the members of $D_1Z_2$ are not independent, and $EZ_1$ unconditionally does not have a normal distribution. So the BLUS residuals

$$R^* = M'Z_1 \quad \text{and} \quad Z_2^* = L_1'Z_2 \tag{42}$$

are constructed, where $M$ is an $n \times n-r-(p-q)$ matrix, $N(M) =$

$\mathcal{M}(E)$, and $\mathbf{M'M} = I_{n-r-p+q}$, and $L_1$ is an $n \times n-r_1$ matrix, $\mathcal{M}(L_1) = \mathcal{M}(D_1)$, and $L_1'L_1 = I_{n-r_1}$. It is straightforward to establish that $R^*$ and $Z_2^*$ in (42) are independent with the following distributions:

$$R^* \sim \mathcal{N}(0, \Gamma_{11 \cdot 2} \otimes I_{n-r-p+q}); \; Z_2^* \sim \mathcal{N}(0, \Gamma_{22} \otimes I_{n-r_1}). \tag{43}$$

As before, $R^*$ is called the set of "between" residuals and $Z_2^*$ the set of "within" residuals.

In terms of a residual analysis, it appears that the most practical approach would be to fit the MANOVA model $\mathbf{Y} = \mathbf{X}_1\boldsymbol{\beta}_1 + \boldsymbol{\varepsilon}$ in the first stage, and from this construct the BLUS residuals from this as $R_1^* = L_1'Y$, where $L_1$ is as defined above. The second stage consists of fitting the GMANOVA model $R_1^* = L_1'X_2\boldsymbol{\beta}_2P + \boldsymbol{\varepsilon}_1^*$. Then the resultant ML estimator of $\boldsymbol{\beta}_2$ with this two-stage process coincides with the ML estimator of $\boldsymbol{\beta}_2$ in (38) under the model in (37).

## 6. NUMERICAL EXAMPLE

Danford, et al. (1960) analyzed the data from a study of 45 patients with cancerous lesions. Patients were assigned to control, low, medium, or high doses of whole-body X-rays according to clinical discretion. Measurements on a psychomotor test were measured at baseline (day 0) and on consecutive days for ten straight days. The clinicians expected the control group to perform poorly over time because the disease would progress unchecked.

Chinchilli and Elswick (1985) used this data set to illustrate the mixed MANOVA-GMANOVA model. The MANOVA portion of the model in (37) contains the covariate of baseline measurement, so that $X_1$ is a $45 \times 1$ matrix and $\boldsymbol{\beta}_1$ is a $1 \times 10$ matrix. The GMANOVA portion of the model contains a distinct linear polynomial curve over time for the different treatment groups, so that $X_2$ is $45 \times 4$, $\boldsymbol{\beta}_2$ is $4 \times 2$, and P is $2 \times 10$. Chinchilli and Elswick (1985) applied their goodness-of-fit tests which revealed that this was an adequate model, and the reader is referred to their article for the details of the analysis.

It is of interest to construct the BLUS residuals from fitting the mixed MANOVA-GMANOVA model to this data set and use them to test normality and homogeneity of variance. The BLUS residual matrices $R^*$ and $Z_2^*$ in (42) are $32 \times 2$ and $44 \times$

8, respectively, and are independent. Applying the multivariate Kolmogorov-Smirnov test, as discussed by Mardia (1980), both the "between" and "within" BLUS residuals reveal that the null hypothesis of normality is not rejected (see Table I).

Unfortunately, it is not possible to apply Bartlett's modified LR test for homogeneity of the variance matrices from the four groups because of inadequate sample sizes (6, 14, 15, and 10 patients in control, low, medium, and high dose groups, respectively). In order to construct a positive definite estimate of the 2 × 2 variance matrix $\Gamma_{11 \cdot 2}$ based on the BLUS residuals within each group, a minimum sample size of twelve is needed for each group. For the 8 × 8 variance matrix $\Gamma_{22}$, a minimum sample size of nine is needed for each group. These follow from the restriction of $n_i$-r-(p-q) ≥ q for the "between" residuals and $n_i$-$r_1$ ≥ p-q for the "within" residuals, where $n_i$ denotes the the sample size of the $i^{th}$ group.

## ACKNOWLEDGEMENTS

Part of this research was supported by the NIH grant AHMD 31370-01A1 from the NIADDK.

Department of Biostatistics, Box 32
Medical College of Virginia
Virginia Commonwealth University
Richmond, Virginia 23298-0001

## REFERENCES

Anscombe, F.J. and Tukey, J.W. (1963). 'The Examination and Evaluation of Residuals,' *Technometrics* **5**, 29-37.

Brown, M.B. and Forsythe, A.B. (1974). 'Robust Tests for the Equality of Variance,' *Journal of the American Statistical Association* **69**, 364-367.

Chinchilli, V.M. and Elswick, R.K. (1985). 'A Mixture of the MANOVA and GMANOVA Models,' *Communications in Statistics - Theory & Methods* **14**, 3075-3089.

Danford, M.B., Hughes, H.M., and McNee, R.C. (1960). 'On the Analysis of Repeated-Measurements Experiments,' *Biometrics* **16**, 547-565.

Das Gupta, S. (1977). 'Two Problems in Multivariate Analysis: BLUS Residuals and Testability of Linear Hypothesis,' *Annals of the Institute of Statistical Mathematics*, **29A**, 35-41.

Freund, R.J., Vail, R.W., and Clunies-Ross, C.W. (1961). 'Residual Analysis,' *Journal of the American Statistical Association* **56**, 98-104.

Grizzle, J.E. and Allen, D.M. (1969). 'Analysis of Growth and Dose Response Curves,' *Biometrics* **25**, 357-381.

Grossman, S.I. and Styan, G.P.H. (1972). 'Uncorrelated Regression Residuals and Singular Values,' *Journal of the American Statistical Association* **67**, 672-673.

Khatri, C.G. (1966). 'A Note on a MANOVA Model applied to problems in growth curves,' *Annals of the Institute of Statistical Mathematics* **18**, 75-86.

Mardia, K.V. (1980). 'Tests of Univariate and Multivariate Normality,' in *Handbook of Statistics, Volume I* (P.R. Krishnaiah, editor). North-Holland: New York, 279-320.

Muirhead, R.J. (1982). *Aspects of Multivariate Statistical Theory*. Wiley: New York.

Potthoff, R.F. and Roy, S.N. (1964). 'A Generalized Multivariate Analysis of Variance Model Useful Especially for Growth Curve Problems,' *Biometrika* **51**, 313-326.

Rao, C.R. (1965). 'Theory of Least Squares When the Parameters Are Stochastic and Its Application to the Analysis of Growth Curves,' *Biometrika* **52**, 447-458.

Rao, C.R. (1967). 'Least Squares Theory Using an Estimated Dispersion Matrix and Its Application to Measurement of Signals,' in *Proceedings of the Fifth Berkeley Symposium on Mathematical Statistics and Probability, Volume I* (L.M. LeCam and J. Neyman, editors). University of California Press: Berkeley, California, 355-372.

Theil, H. (1965). 'The Analysis of Disturbances in Regression Analysis,' *Journal of the American Statistical Association* **60**, 1067-1079.

Theil, H. (1968). 'A Simplification of the BLUS Procedure for analyzing regression disturbances,' *Journal of the American Statistical Association* **63**, 242-251.

Zyskind, G. (1963). 'A Note on Residual Analysis,' *Journal of the American Statistical Association* **58**, 1125-1132.

Table I

The multivariate Kolmogorov-Smirnov tests of normality applied to the "between" and "within" BLUS residuals of the Danford, et al. (1960) data.

| BLUS Residuals | K-S Test Statistic | P-Value |
|---|---|---|
| Between | 0.479463 | 0.9756 |
| Within | 0.998774 | 0.2713 |

Sung C. Choi and Vernon M. Chinchilli

# ANALYSIS OF WITHIN- AND ACROSS-SUBJECT CORRELATIONS

## ABSTRACT

In some fields of applications, response variables are measured on $k(k > 1)$ independent samples for each experimental subject. For this type of situation, within-subject and across-subject correlation matrices are defined and methods of analysis are discussed.

The maximum likelihood estimators for the two different correlation matrices are obtained, and the exact test for within-subject correlation and two approximate tests for across-subject correlation are proposed.

Simulation studies for bivariate distributions suggest that the estimators are satisfactory although the across-subject correlation coefficients are somewhat under estimated. The studies also showed that the two approximate tests are adequate in terms of the size and power. Other properties of the estimators and the tests are discussed.

**Key words and phrases:**
Multivariate Model, Within-Subject Correlation Matrix, Across-Subject Correlation Matrix, Estimation, Testing, Computer Simulation

## 1. INTRODUCTION

Suppose that there are n subjects in a study and that p response variables are measured on k independent samples (e.g., locations or occasions) in each subject. Suppose that the correlation between the p variables is of interest. For example, in head injury research, investigators are frequently concerned with the correlation between the levels of different chemical substances in randomly selected tissues of the brain.



*H. Bozdogan and A. K. Gupta (eds.), Multivariate Statistical Modeling and Data Analysis, 77–93.*

For this type of situation, often two different forms of correlation matrices must be dealt with: the within-subject correlation matrix and the across-subject correlation matrix. Roughly, the former is the correlation between variables within a typical subject while the latter is the correlation between the variable means as they vary across subjects. From a statistical point of view, the former may be more conveniently defined as the correlation matrix of error terms. In this paper, however, we shall call it the within-subject correlation matrix because in many practical problems the use of the term "error" seems inappropriate. The two different correlation matrices are more precisely defined in the next section. The purpose of this paper is to study the problem of estimating and testing hypothesis about the two correlation matrices.

The related estimation problem when p = 2, all from a regression point of view, has been studied by several authors including Wald (1940), Lindley (1947), Grubbs (1947), Geary (1949), Reiersal (1950), Tukey (1951) and Healy (1958). It would appear, however, that properties of estimators for finite n and k have not been studied. In addition, little work has been done on testing hypotheses about the two different correlation matrices, especially for the across-subject correlation matrix.

## 2. DEFINITIONS AND MODELS

Let

$$\underset{\sim}{Z}'_{ij} = \left[ Z^{(1)}_{ij} \; \ldots \; Z^{(p)}_{ij} \right] \tag{1}$$

denote the p-vector of responses on the jth location for the ith subject, $1 \leq i \leq n$ and $1 \leq j \leq k$. The model postulated for $\underset{\sim}{Z}_{ij}$ is

$$\underset{\sim}{Z}'_{ij} = \underset{\sim}{a}'_i \underset{\sim}{B} + \underset{\sim}{U}'_i + \underset{\sim}{\varepsilon}'_{ij} \quad , \tag{2}$$

where $\underset{\sim}{a}'_i$ is a row vector of r fixed covariates, $\underset{\sim}{B}$ is a r-by-p unknown parameter matrix, $\underset{\sim}{U}'_i$ is a row vector of p random components representing the mean effect of the ith subject, and $\varepsilon'_{ij}$ is a row vector of p random components representing the effect of the jth measurement on the ith subject. This model may be considered as essentially a mixed model in multivariate regression with r fixed effects and one random effect each with p components. It is assumed that all $U_i$ and $\varepsilon_{ij}$ have independent p-variate normal distributions:

$$U_i \sim N(0, \Omega), \qquad \varepsilon_{ij} \sim N(0, \Gamma)$$

We shall call $\Omega$ the across-subject covariance matrix and $\Gamma$ the within-subject covariance matrix. The two different forms of correlation matrices are analogously defined.

Let $Z'_i = [Z'_{i1} \cdots Z'_{ik}]$ denote the row vector of all pk observations on the ith subject. The model given by (1) for $Z'_i$ can be written as

$$Z' = a'_i\left(j \otimes B\right)+\left(j \otimes U'_i\right)+\varepsilon'_i \qquad (3)$$

where $j$ is a row vector with k unit values and $\varepsilon'_i = [\varepsilon'_{i1} \cdots \varepsilon'_{ik}]$. Throughout the paper we shall represent the direct (Kronecker) product of $j$ and $B$, for example, by $j \otimes B$. The covariance matrix of $Z_i$ denoted by $\Delta$, is shown to be

$$\Delta = \left(J_k \otimes \Omega\right)+\left(I_k \otimes \Gamma\right) , \qquad (4)$$

where $J_k$ is the k-by-k matrix with unit value for all elements and $I_k$ is the k-by-k identity matrix.

Finally, the model for all n subjects can be constructed using the following matrices:

$$Z = \begin{bmatrix} Z'_1 \\ \vdots \\ Z'_n \end{bmatrix}, \quad A = \begin{bmatrix} a'_1 \\ \vdots \\ a'_n \end{bmatrix}, \quad U = \begin{bmatrix} U'_1 \\ \vdots \\ U'_n \end{bmatrix}, \quad \text{and } \varepsilon = \begin{bmatrix} \varepsilon'_1 \\ \vdots \\ \varepsilon'_n \end{bmatrix} .$$

Without loss of generality, it is assumed that $\underset{\sim}{A}$ is of full rank r. Then the model can be written as

$$\underset{\sim}{Z} = \underset{\sim}{A}(\underset{\sim}{j} \otimes \underset{\sim}{B})+(\underset{\sim}{j} \otimes \underset{\sim}{U})+\underset{\sim}{\varepsilon} \quad . \tag{5}$$

In addition to the matrices defined so far, the following three are also used throughout this paper.

$$\underset{\sim}{Z}_{i.} = k^{-1} \sum_{j=1}^{k} \underset{\sim}{Z}_{ij}, \quad 1 \leq i \leq n \quad ,$$

$$\underset{\sim}{S}_1 = \sum_{i=1}^{n} \sum_{j=1}^{k} \left(\underset{\sim}{Z}_{ij}-\underset{\sim}{Z}_{i.}\right)\left(Z_{ij}-Z_{i.}\right)' \quad , \tag{6}$$

$$\underset{\sim}{S}_2 = k\, \overline{\underset{\sim}{Z}}'\left\{\underset{\sim}{I}_n-\underset{\sim}{A}(\underset{\sim}{A}'\underset{\sim}{A})^{-1}\underset{\sim}{A}'\right\}\overline{\underset{\sim}{Z}} \, , \tag{7}$$

where $\overline{\underset{\sim}{Z}} = \left[\underset{\sim}{Z}_{1.}-\underset{\sim}{Z}_{..} \cdots \underset{\sim}{Z}_{n.}-\underset{\sim}{Z}_{..}\right]'$ and $\underset{\sim}{Z}_{..} = \Sigma\, \underset{\sim}{Z}_{i.}/n$.

## 3. ESTIMATION OF CORRELATION MATRICES AND COEFFICIENTS

In this section we shall be briefly concerned with the maximum likelihood estimation (MLE) of $\underset{\sim}{\Omega}$ and $\underset{\sim}{\Gamma}$.

The log likelihood function for $\underset{\sim}{Z}$ is

$$L = C+(n/2)\log\left|\Delta^{-1}\right|-(1/2)\sum^{n} \left\{\underset{\sim}{Z}_i'-\underset{\sim}{a}_i'(\underset{\sim}{j} \otimes \underset{\sim}{B})\right\}\Delta^{-1}\left\{\underset{\sim}{Z}_i-(\underset{\sim}{j} \otimes \underset{\sim}{B})'\underset{\sim}{a}_i\right\}, \tag{8}$$

where C is a constant which is independent of the parameters. The usual differentiation procedure along with some algebraic manipulations yield the following MLE of $\underset{\sim}{\Gamma}$ and $\underset{\sim}{\Omega}$

$$\hat{\underset{\sim}{\Gamma}} = \left\{n(k-1)\right\}^{-1}\underset{\sim}{S}_1 \quad , \tag{9}$$

$$\hat{\Omega} = k^{-1}\left\{(n-r)^{-1}S_2-\hat{\Sigma}\right\} . \tag{10}$$

Note that the divisors n(k-1) and n-r have been used in (9) and (10), respectively, in order for $\hat{\Sigma}$ and $\hat{\Omega}$ to be unbiased. although $\hat{\Omega}$ is unbiased, it has positive probability of being an indefinite matrix. Bock and Petersen (1975) provided a smoothing technique in which the smoothed estimator is at least nonnegative definite with probability one.

In particular, consider the within-subject and the across-subject correlation coefficients for two selected variables X and Y, denoted simply as $\rho_W$ and $\rho_b$, respectively. It is convenient to introduce the usual notations for the sum of squares and cross-products terms as in Table I. When p = 2, (9) and (10) yield the following fairly well-known MLE of $\rho_W$ and the relatively unknown and uninvestigated MLE of $\rho_b$:

$$\hat{\rho}_W = \frac{E_{xy}}{\left[E_{xx}E_{yy}\right]^{1/2}} , \tag{11}$$

$$\hat{\rho}_b = \frac{T_{xy}/(n-r)-E_{xy}/(N-n)}{\left[\left\{T_{xx}/(n-r)-E_{xx}/(N-n)\right\}\left\{T_{yy}/(n-r)-E_{yy}/(N-n)\right\}\right]^{1/2}} , \tag{12}$$

where N = nk.

We could consider two other estimators of $\rho_W$ each based on n independent estimators $\hat{\rho}_W$ (i = 1, ... n) where $\hat{\rho}_W$ is the usual sample correlation coefficient calculated from k observations for the ith individual. The first estimator is the sample mean of the n coefficients $\hat{\rho}_W$. The second is the pooled coefficient based on the fact that $(k-3)^{1/2}[\text{arctanh}(\hat{\rho}_W)-\text{arctanh}(\rho_W)]$ converges in law to the N(0, 1) as $k \to \infty$. It is not surprising to find that these two estimators are not as efficient as (11), at least for the nominal size of k according to a simulation study of Section 5.

In the derivation of the MLE for $\Omega$, if k is not constant over individuals, we could not expect to find a closed form for the MLE of $\rho_b$. However, assume that k is a random variable. The assumption enables us to obtain a reasonable ad-hoc estimator of $\rho_b$ using the definition of $\rho_b$. Let $\rho_b\sigma_1\sigma_2$ and $\rho_w\gamma_1\gamma_2$ be the off-diagonal elements of $\Omega$ and $\Gamma$, respectively.

Consider the unweighted means $(X_{i.}, Y_{i.})$ disregarding the difference between k's. Using the relation, $\mathrm{var}(X.) = E\{\mathrm{var}(X.|k)\}+\mathrm{var}\{E(X.|k)\}$ we have

$$\mathrm{var}(X.) = \sigma_1^2+\gamma_1^2 E(1/k), \tag{13}$$

and a similar expression for var(Y.). Also, from the equality $\mathrm{cov}(X., Y.) = E\{\mathrm{cov}(X., Y.)|k\}+\mathrm{cov}\{E(X.|k), E(Y.|k)\}$, we get

$$\mathrm{cov}(X., Y.) = \rho_b\sigma_1\sigma_2+\rho_w\gamma_1\gamma_2\ E(1/k) \tag{14}$$

From (13) and (14), we obtain the following estimator of the correlation coefficient, $\rho_b$, based on the various estimators given in the section along with the reciprocal of the harmonic mean as the estimator for E(1/k):

$$\rho_b = \frac{T_{xy}/(n-r)-E_{xy}/nH}{\left[\left\{T_{xx}/(n-r)-E_{xx}/nH\right\}\left\{T_{yy}/(n-r)-E_{yy}/nH\right\}\right]^{1/2}}, \tag{15}$$

where H represents the harmonic mean of $k_1, k_2, \ldots, k_n$.

## 4. TEST OF CORRELATION MATRICES AND COEFFICIENTS

### 4.1 Test of Within-Subject Correlation Matrix and Coefficient

Little work has been done on testing the two types of correlation matrices, especially for $\Omega$. First, inference on the parameter matrix $\Gamma$ is not difficult because $n(k-1)\hat{\Gamma} = S_1$ is a Wishart matrix with n(k-1) degrees of freedom. It follows from Anderson ((1984), Sec. 10.8) that the likelihood ratio (LR) criterion for testing $H_0$: $\Gamma = \Gamma_0$, a specified matrix, is

$$\lambda = \left|S_1\Gamma_0^{-1}\right|^{nk/2}\exp\left[-\left\{\text{trace}\left[S_1\Gamma_0^{-1}\right]-pnk\left[1-\log(nk)\right]\right\}\Big/2\right] .$$

Thus, we can use the test statistic $-2\log\lambda$ which is asymptotically distributed as a chi-square distribution with $p(p+1)/2$ degrees of freedom.

For practical applications, testing $H_0$: $\Gamma = \Gamma_0$ may not be as important as testing that the correlation coefficients are null. Let

$$R_w = \left\{\text{Diag}^{-1/2}(\Gamma)\right\}\Gamma\left\{\text{Diag}^{-1/2}(\Gamma)\right\}$$

represent the correlation matrix associated with $\Gamma$, and similarly let

$$\hat{R}_w = \left\{\text{Diag}^{-1/2}(\hat{\Gamma})\right\}\hat{\Gamma}\left\{\text{Diag}^{-1/2}(\hat{\Gamma})\right\} \quad (16)$$

represent the corresponding sample correlation matrix defined from $\hat{\Gamma}$. Anderson ((1984), Sec. 7.6) has shown that the set of sample correlation coefficients in $\hat{R}_w$ has the following exact density under $H_0$: $R_w = I_p$

$$\frac{G^p\left[n(k-1)/2\right]\left|\hat{R}_w\right|^{\{n(k-1)-p-1\}/2}}{\pi^{p(p-1)/4}\prod_{j=1}^{p} G[\{n(k-1)+1-j\}/2]}, \quad (17)$$

where $G(k)$ is the gamma function of $k$. In particular, for testing $H_0$: $\rho_w = 0$ when $p = 2$, (17) reduces to the well-known test statistics with a t-distribution.

## 4.2 Test of Across-Subject Correlation Matrix and Coefficient

The problem of testing hypotheses about the parameter matrix $\Omega$ is somewhat complicated and does not appear to be dealt with in the literature. Note that the estimator of $\Omega$ given by (10) is not distributed as a Wishart form.

Consider the problem of testing the null hypothesis $H_0$: $\Omega = \Omega_0$, a specified matrix. In order to construct the LR test of this hypothesis, the maximum likelihood estimator of $\Gamma$ under the null hypothesis must be found. Unfortunately, this requires an iterative solution of the following matrix equation with respect to $\Gamma$

$$n(k-1)\Gamma - S_1 + \Gamma\left(\Gamma + k\Omega_0\right)^{-1}\left[n\left(\Gamma + k\Omega_0\right) - S_2\right]\left(\Gamma + k\Omega_0\right)^{-1}\Gamma = 0.$$

The solution of this equation and in particular a relevant simulation study of the LR would be not very practical. Even when $p = 2$, one must deal with three simultaneous equations each with a polynomial function of degree three. Another criterion, the score statistic proposed by Rao (1947), also requires a similar solution. An alternative approach to testing $H_0$: $\Omega = \Omega_0$ is Wald's statistic (1943) which is asymptotically equivalent to the LR test. The advantage of this statistic is that it does not require the explicit computation of the MLE under $H_0$.

Wald's test statistic for $H_0$ is derived as follows. First, the p-by-p score matrix for $\Omega$ is

$$S(\Omega) = \left(k/2\right)\left(\Gamma + k\Omega\right)^{-1}\left[S_3 - n\left(\Gamma + k\Omega\right)\right]\left(\Gamma + k\Omega\right)^{-1}, \quad (18)$$

where

$$S_3 = k(\bar{Z} - AB)'(\bar{Z} - AB) \quad .$$

Using (18). the $p^2$-by-$p^2$ information matrix for $\Omega$ can be obtained as

$$I(\Omega) = \left(nk^2/4\right)\left[\left(\Gamma + k\Omega\right)^{-1} \otimes \left(\Gamma + k\Omega\right)^{-1}\right]\left(I_{p^2} + I_{(p,p)}\right) , \quad (19)$$

where $\underset{\sim}{I}_{(p,p)}$ is the $p^2$-by-$p^2$ commutation matrix (see Magnus and Neudecker, (1979)).

Let $\underset{\sim}{W}$ and $W_0$ denote the vectors of the p(p+1)/2 unique elements of $\underset{\sim}{\Omega}$ and $\underset{\sim}{\Omega}_0$, respectively. Analogous to this, let $\underset{\sim}{I}(\underset{\sim}{W})$ denote the p(p+1)/2 by p(p+1)/2 information matrix for $\underset{\sim}{W}$ given by the submatrix of $\underset{\sim}{I}(\underset{\sim}{\Omega})$, deleting the corresponding rows and columns. Since the appropriate regularity conditions are satisfied, it follows that $n^{1/2}[\hat{\underset{\sim}{W}}-\underset{\sim}{W}_0]$ converges in law to $N[\underset{\sim}{0}, n\underset{\sim}{I}^{-1}(\underset{\sim}{W})]$. A Wald type construction for testing $H_0$: $\underset{\sim}{\Omega} = \underset{\sim}{\Omega}_0$ leads to the following statistic which has an asymptotic chi-square distribution with p(p+1)/2 degrees of freedom:

$$\left[\hat{\underset{\sim}{W}}-\underset{\sim}{W}_0\right]'\underset{\sim}{I}(\hat{\underset{\sim}{W}})\left[\hat{\underset{\sim}{W}}-\underset{\sim}{W}_0\right] , \tag{20}$$

where $\underset{\sim}{I}(\hat{\underset{\sim}{W}})$ is given by $\underset{\sim}{I}(\underset{\sim}{W})$ with $\hat{\underset{\sim}{W}}$ defined from (10) substituted for $\underset{\sim}{W}$. In order to construct a test for the correlations matrix defined from $\underset{\sim}{\Omega}$, let

$$\underset{\sim}{R}_b = \left\{\mathrm{Diag}^{-1/2}(\underset{\sim}{\Omega})\right\} \underset{\sim}{\Omega} \left\{\mathrm{Diag}^{-1/2}(\underset{\sim}{\Omega})\right\}$$

be the across-subject correlation matrix. Then Wald's construction of an asymptotic test for $H_0$: $\underset{\sim}{R}_b = \underset{\sim}{I}_{p(p-1)/2}$ can be found in a similar manner. This is because $H_0$: $\underset{\sim}{R}_b = \underset{\sim}{I}_{p(p-1)/2}$ is equivalent to $H_0$: $\underset{\sim}{\Omega}$ is diagonal.

Let $\hat{\underset{\sim}{V}}$ denote the vector of p(p-1)/2 unique off-diagonal elements of $\hat{\underset{\sim}{\Omega}}$, and let $\underset{\sim}{I}^{-1}(\underset{\sim}{V})$ denote the corresponding p(p-1)/2 by p(p-1)/2 submatrix of $\underset{\sim}{I}^{-1}(\underset{\sim}{W})$. Then, as for (20), the statistic for testing $H_0$: $\underset{\sim}{R}_b = \underset{\sim}{I}_{p(p-1)/2}$ can be obtained as

$$T_1 = \hat{\underset{\sim}{V}}'\underset{\sim}{I}(\hat{\underset{\sim}{V}})\underset{\sim}{V} , \tag{21}$$

which has an asymptotic chi-square distribution with p(p-1)/2 degrees of freedom. The test $T_1$ is asymptotically equivalent to the LR test.

As a special case of p = 2, consider the problem of testing that the across-subject correlation coefficient $\rho_b$ between X and Y is zero. The test statistic (21) can be expressed using the terms of Table I in the following form which is asymptotically distributed as a chi-square with a single degree of freedom:

$$T_1 = \frac{\left\{T_{xy}/(n-r) - E_{xy}/(N-n)\right\}^2}{T_{xx}T_{yy}/n(n-r)^2} \tag{22}$$

Another test for $H_0$: $\rho_b = 0$ can be derived as follows. First, the expectation of the numerator of (12) can be shown to be $k\rho_b\sigma_x\sigma_y$, and hence it is zero under $H_0$. Moreover, $T_{xy}$ and $E_{xy}$ are not only asymptotically normal but are independent. Next, as in Tukey's (1951) approach in regression, the variance of the numerator can be approximated by the conditional variance assuming that X is fixed, if n and k are not too small. An ad-hoc approximate test can be obtained as

$$T_2 = \frac{\left\{T_{xy}/(n-r) - E_{xy}/(N-n)\right\}^2}{T_{xx}T_{yy}/n(n-r)^3 + E_{xx}E_{yy}/(N-n)^3}, \tag{23}$$

which has asymptotically a chi-square distribution with a single degree of freedom. The above two tests are asymptotically equivalent.

## 5. SIMULATION STUDIES

The property of the estimators given by (11), (12) and (15) and comparative efficiencies of the two tests given by (22) and (23) were examined by simulation experiments. Briefly, a normal bivariate random number generator was used to generate a set of n bivariate random numbers $\underset{\sim}{U}_i$ with a desired covariance matrix $\underset{\sim}{\Omega}$ and zero mean vector. Next, for each $\underset{\sim}{U}_i$ the second normal bivariate random number generator was used to generate k independent $\underset{\sim}{\varepsilon}_{ij}$ with a desired covariance matrix $\underset{\sim}{\Sigma}$ and zero mean vector. These two processes are combined to generate $\underset{\sim}{Z}_{ij}$. Without loss of generality, $\mu$ is taken to be zero assuming r = 1.

The first experiment was performed to investigate estimators $\hat{\rho}_w$ given by (11) and $\hat{\rho}_b$ given by (12). Sample size was chosen to be 30 for n, and k was fixed at 5. $\sigma_1^2$

Table I

Sums of squares and products terms when p = 2

| Source of Variation | Degrees of Freedom | For X | For XY | For Y |
|---|---|---|---|---|
| Across Subject | n-1 | $\sum k_i(X_{i.}-X_{..})^2 = T_{xx}$ | $\sum k_i(X_{i.}-X_{..})(Y_{i.}-Y_{..}) = T_{xy}$ | $\sum k_i(Y_{i.}-Y_{..})^2 = T_{yy}$ |
| Within Subject | N-n | $\sum\sum(X_{ij}-X_{i.})^2 = E_{xx}$ | $\sum\sum(X_{ij}-X_{i.})(Y_{ij}-Y_{i.}) = E_{xy}$ | $\sum\sum(Y_{ij}-Y_{i.})^2 = E_{yy}$ |

Table II

Mean value of across- and within-subject correlations, $\hat{\rho}_b$ and $\hat{\rho}_w$ for given $\sigma_1^2$, $\rho_b$ and $\rho_w$ based on simulation when n = 30

| $\sigma_1^2$ | $\rho_w$ | $\rho_b$ | $\hat{\rho}_b$ | $\hat{\rho}_w$ |
|---|---|---|---|---|
| | | | k = 5 | |
| 5.0 | 0.9 | 0.3 | 0.290 | 0.900 |
| | | 0.1 | 0.095 | 0.900 |
| | | 0.0 | -0.001 | 0.900 |
| | | -0.3 | -0.301 | 0.899 |
| | 0.6 | 0.3 | 0.284 | 0.596 |
| | | 0.1 | 0.102 | 0.597 |
| | | 0.0 | -0.011 | 0.600 |
| | | -0.3 | -0.306 | 0.597 |
| 0.5 | 0.9 | 0.3 | 0.265 | 0.900 |
| | | 0.1 | 0.052 | 0.899 |
| | | 0.0 | -0.042 | 0.900 |
| | | -0.3 | -0.379 | 0.900 |
| | 0.6 | 0.3 | 0.274 | 0.597 |
| | | 0.1 | 0.081 | 0.597 |
| | | 0.0 | -0.043 | 0.597 |
| | | -0.3 | -0.333 | 0.601 |
| | | | Variable k | |
| 5.0 | 0.9 | 0.3 | 0.294 | 0.899 |
| | | 0.1 | 0.098 | 0.899 |
| | | 0.0 | -0.003 | 0.899 |
| | | -0.3 | -0.299 | 0.900 |
| | 0.6 | 0.3 | 0.300 | 0.599 |
| | | 0.1 | 0.101 | 0.602 |
| | | 0.0 | -0.003 | 0.597 |
| | | -0.3 | -0.293 | 0.599 |

and $\sigma_2^2$ were variably fixed at 5.0, 1.0 and 0.5, while $\gamma_1^2$ and $\gamma_2^2$ were fixed at 1.0. The correlation coefficients were selected as follows: $\rho_b$ = 0.3, 0.1, 0.0 and -0.3, and $\rho_w$ = 0.9 and 0.6. For each combination of the four parameters, 1,000 independently simulated experiments were used.

The result of the first simulation study is presented in Table II which gives the mean of 1,000 empirical estimates for each $\hat{\rho}_b$, and $\hat{\rho}_w$. As expected, $\hat{\rho}_w$, is very efficient; $\hat{\rho}_w$ is essentially the MLE of $\rho_w$ based on a sample of 150 independent observations. Also, as expected, $\hat{\rho}_w$ is a slight underestimator of $\rho_w$ since we know that its negative bias is of O(1/n). (Olkin and Pratt (1958) and Muirhead (1982) discuss the unbiased version of $\hat{\rho}_w$, but it involves an infinite series.) As mentioned in Section 3, the two other possible estimators for $\rho_w$ were found to be less efficient than $\hat{\rho}_w$. Of course, our primary interest centers on $\hat{\rho}_b$. Table I shows that $\hat{\rho}_b$ based on (12) is a reasonable estimator although it tends to underestimate $\rho_b$ somewhat. Roughly, the bias appears to be increasing as $\rho_w$ increases relative to $\rho_b$ and/or $\sigma_1^2$ and $\sigma_2^2$ decrease relative to $\gamma_1^2$ and $\gamma_2^2$. The standard error of $\hat{\rho}_b$ in the table ranged from 0.006 to 0.009.

The second simulation experiment was conducted to assess the estimator given by (15) when k is not constant. Again, n was fixed to 30 but k was chosen according to the following multinomial probability P(k): P(3) = P(7) = 0.1, P(4) = P(6) = 0.2 and P(5) = 0.4. The same process used in the first simulation study was employed except that k was determined by a multinomial random number generater. Again, the same combinations of the parameters were used with 1,000 independent simulation results for each combination of the parameters. A part of the result of the experiment is presented also in Table I. The overall property including the accuracy and precision of the estimator was just about the same as that when k is fixed.

The third simulation study was performed to assess and compare the two tests for $\rho_b$ proposed in Section 4. The nominal significance level was set to $\alpha$ = 0.05 and $\rho_b$ was variably fixed at -0.6, 0.0, 0.3 and 0.6 with various values of $\rho_w$. Sample size ranging from 15 to 50 for n was considered with several values of k. For each combination of the parameters and n and k, 4,000 independent experiments were

Table III

Empirical power (in %) of the two proposed tests for $\rho_b = 0$ under various combinations of parameters when $\gamma_1^2 = 1$ and $k = 5$

| $\rho_w$ | $\gamma_2^2$ | $\sigma_1^2$ | $\sigma_2^2$ | $\rho_b$ | n = 15 $T_1$ | n = 15 $T_2$ | n = 30 $T_1$ | n = 30 $T_2$ |
|---|---|---|---|---|---|---|---|---|
| | | | | -0.6 | 61.8 | 57.7 | 90.6 | 89.6 |
| | 1 | 2 | 2 | 0.0 | 5.8 | 4.2 | 5.6 | 4.0 |
| | | | | 0.3 | 16.0 | 14.8 | 32.8 | 31.0 |
| | | | | 0.6 | 59.8 | 57.3 | 90.3 | 88.8 |
| | | | | -0.6 | 69.3 | 66.0 | 94.8 | 94.3 |
| 0.2 | 1 | 10 | 10 | 0.0 | 4.9 | 4.5 | 5.1 | 5.2 |
| | | | | 0.3 | 20.8 | 17.7 | 36.8 | 35.0 |
| | | | | 0.6 | 68.6 | 65.2 | 95.5 | 94.7 |
| | | | | -0.6 | 63.0 | 58.6 | 90.6 | 90.3 |
| | 2 | 2 | 10 | 0.0 | 5.6 | 4.7 | 5.3 | 5.1 |
| | | | | 0.3 | 17.0 | 15.8 | 32.3 | 31.8 |
| | | | | 0.6 | 60.1 | 59.9 | 91.9 | 91.2 |
| | | | | -0.6 | 60.0 | 58.0 | 90.5 | 89.0 |
| | 1 | 2 | 2 | 0.0 | 6.3 | 4.9 | 5.6 | 4.8 |
| | | | | 0.3 | 16.5 | 14.4 | 31.0 | 30.6 |
| | | | | 0.6 | 61.8 | 57.2 | 90.7 | 89.5 |
| | | | | -0.6 | 69.3 | 65.7 | 94.4 | 94.0 |
| 0.8 | 1 | 10 | 10 | 0.0 | 5.8 | 4.5 | 5.3 | 5.2 |
| | | | | 0.3 | 19.1 | 16.6 | 36.3 | 34.7 |
| | | | | 0.6 | 70.1 | 66.0 | 93.8 | 93.4 |
| | | | | -0.6 | 64.4 | 61.1 | 91.6 | 90.6 |
| | 2 | 2 | 10 | 0.0 | 5.3 | 5.1 | 5.2 | 4.2 |
| | | | | 0.3 | 16.6 | 16.0 | 33.4 | 31.1 |
| | | | | 0.6 | 63.2 | 60.9 | 93.5 | 92.5 |

used, and in each experiment both tests were applied. The empirical power of the two tests when $k = 5$ is reproduced in Table III in part.

A rather clear picture emerged from the study and the following conclusions can be drawn about the tests $T_1$ and $T_2$ given by (22) and (23), respectively: (i) The size of $T_1$ tends to be somewhat greater than the nominal value unless n is large, $(n > 30)$. The size of $T_2$ appears to be satisfactory regardless of n. (ii) Taking the difference in the size into account, the efficiencies of $T_1$ and $T_2$ appear to be roughly comparable. (iii) Other parameters being the same, power of each test tends to increase with increasing $\sigma_1^2$ and $\sigma_2^2$ relative to $\gamma_1^2$ and $\gamma_2^2$. (iv) To some surprise, the effect of $\rho_w$ on the power function is almost nil in both tests. (v) For a given n the power of each test increases with increasing k. For $k = 2$ the power was about 10 to 20 percent lower than when $k = 5$.

## 6. CONCLUSIONS

The method of estimating the testing both within- and across-subject correlation coefficients are examined and simulation studies were carried out. The results indicated that the maximum likelihood estimator for the across-subject correlation coefficient $\hat{\rho}_b$ is reasonable although it tends to slightly underestimate $\rho_b$. For testing hypotheses about $\rho_b$, we recommend $T_2$ on the basis of its better behavior in terms of the power function and its computational simplicity. However, for testing hypotheses about several across-subject correlations simultaneously when $p > 2$ or the correlation matrix, only $T_1$ is practical.

ACKNOWLEDGEMENT

This research was supported in part by NINCDS Grant NS-12587 from the National Institutes of Health.

Department of Biostatistics
Medical College of Virginia
Virginia Commonwealth University
P.O. Box 32, MCV Station
Richmond, VA 23298-0001

## REFERENCES

T. W. Anderson (1984). An Introduction to Multivariate Statistical Analysis. Wiley, New York.

M. S. Bartlett (1950). 'Fitting a Straight Line When Both Variables are Subject to Error' Biometrics, **5**, 207-212.

J. Berkson (1950). 'Are there two regressions?' J. Am. Statist. Assoc., **45**, 164-180.

R. D. Bock and A. C. Petersen (1975). 'A Multivariate Correction for Attenuation' Biometrika, **62**, 673-678.

W. G. Cochran (1968). 'Errors of Measurement in Statistics' Technometrics, **10**, 637-666.

J. Durbin (1954). 'Errors in Variables' Rev. Inter. Statist. Inst. **22**, 23-32 (1954).

R. C. Geary, R.C. (1949). 'Determination of Linear Relations Between Systematic Parts of Variables With Errors of Observation the Variance of Which are Unknown' Econometrika, **17**, 30-58.

F. E. Grubbs (1948). 'On Estimating Precision of Measuring Instruments and Product Variability' J. Am. Statist. Assoc., **43**, 243-264.

M. J. R. Healy (1958). 'Variations Within Individuals in Human Biology' Human Biology, **30**, 210-218.

L. Kish (1962). 'Studies of Interviewer Variance for Attitudinal Variables' J. Am. Statist. Assoc. **57**, 92-115.

D. V. Lindley (1947). 'Regression Lines and the Linear Functional Relationship' J. Roy. Statist. Soc. Suppl. **9**, 218-244.

A. Madansky (1959). 'The Fitting of Straight Lines When Both Variables are Subject to Error' J. Am. Statist. Assoc., **54**, 173-205.

J. R. Magnus and H. Neudecker (1979). 'The Commutation matrix: Some properties and Applications' Annals of Statistics, 7, 381-394.

R. J. Muirhead (1982). Aspects of Statistical Multivariate Theory, page 157. New York: John Wiley and Sons.

I. Olkin and J. W. Pratt (1958). Unbiased Estimation of Certain Correlation Coefficients. Annals of Mathematical Statistics, **29**, 201-211.

S. J. Press (1979). 'Matrix Intraclass Covariance Matrices With Applications in Agriculture' Technical Report No. 49, Department of Statistics, University of California, Riverside.

C. R. Rao (1947). 'Large Sample Tests of Statistical Hypotheses Concerning Several Parameters With Applications to Problems of Estimation' Proc. Camb. Phil. Soc. **44**, 50-57.

O. Reiersal (1950). 'Identifiability of a Linear Relation Between Variables Which are Subject to Error' Econometrika, **18**, 375-389.

J. W. Tukey (1951). 'Components in Regression' Biometrics, **7**, 33-69.

A. Wald (1940). 'The Fitting of Straight Lines if Both Variables are Subject to Error' Annals Math. Statist., **11**, 284-300.

A. Wald (1943). 'Test of Statistical Hypothesis Concerning Several Parameters When the Number of Observations is large. Trans. Am. Math. Soc., **54**, 426-482.

Dorothea Eisenblätter and Hamparsum Bozdogan

# TWO-STAGE MULTI-SAMPLE CLUSTER ANALYSIS AS A GENERAL APPROACH TO DISCRIMINANT ANALYSIS

## ABSTRACT

This paper introduces *Two-Stage Multi-Sample Cluster Analysis* (TSMSCA), i.e., the problem of grouping samples and improving upon homogeneity via reassigning individual objects, as a general approach to 'classical' discriminant analysis (DA).

Akaike's Information Criterion (AIC) and Bozdogan's CAIC are derived and used in TSMSCA to choose the best fitting model and the best partition among all possible clustering alternatives. With this approach the dimension of the discriminant space is determined, and using a decision-tree classifier, the best lower dimensional models are identified, yielding a hierarchy of efficient separation and assignment rules. On each step of the hierarchy, the performance of the classification of the best discriminant model is evaluated either by a cross-validation method or the method of conditional clustering.

Cross-validation reassigns one object at a time based only on the tentatively updated model, whereas the conditional clustering method actually executes reassignments of objects via a transfer and swapping algorithm given the best discriminant model as the initial partition.

Numerical examples are carried out on real data sets to demonstrate the generality and versatility of the proposed new approach.

Key words and phrases: Two-Stage Multi-Sample Cluster Analysis; Cluster Analysis; Discriminant Analysis; AIC; CAIC.

## 1. INTRODUCTION

Many practical situations require the assignment of individual elements of unknown origin to one of two or more categories on the basis of the values of several characteristics. Traditionally,



*H. Bozdogan and A. K. Gupta (eds.), Multivariate Statistical Modeling and Data Analysis, 95–119.*

the objective of discriminant analysis is to construct an assignment rule using available data, typically from *training samples* of categorized observation units.

An assignment rule is usually assessed by estimates of misclassification error probabilities, called *error rates*, computed from the set of training samples. Prior to constructing an assignment rule the researcher has to determine the appropriate parametric model and the dimension of the discriminant space. Considering the collection of decisions to be made in and prior to *'classical' discriminant analysis* (DA), it would certainly be desirable to apply a method, which selects and evaluates the best fitting model among all possible alternatives without subjective interference.

The purpose of this paper is, therefore, to propose and study *Two-Stage Multi-Sample Cluster Analysis* (TSMSCA), the problem of grouping samples and improving upon homogeneity via reassigning individual elements, as a general approach to 'classical' discriminant analysis.

Akaike's Information Criterion (AIC), due to Akaike (1973), (1974), and Bozdogan's CAIC [Bozdogan (1987)] are used in TSMSCA to choose the best fitting parametric model and the best clustering alternative among all possible alternatives. With this approach the dimension of the discriminant space can be determined, and using a decision-tree classifier the best lower dimensional models can be identified yielding a hierarchy of separation and assignment rules. Furthermore, on each step of the hierarchy, we can evaluate the performance of the best assignment rule of the respective dimension either by a cross-validation method or the method of conditional clustering.

In Section 2, we shall briefly discuss DA, sketch its formulation, and point out existing problems. In Section 3, we shall describe the general TSMSCA procedure. We shall define the general TSMSCA problem, present *Multi-Sample Cluster Analysis* (MSCA) [see Bozdogan (1983), Bozdogan and Sclove (1984)], the problem of clustering samples, as the first stage, and the problem of improving upon homogeneity via reassigning individual elements as the second stage of TSMSCA. Subsequently, in Section 4, we shall present the formulation of model selection and evaluation in TSMSCA. We shall briefly present the model-selection criteria, and explain how they are generally employed in a decision-making process for discriminant analysis. In Section 5, we shall carry out numerical examples on two real data sets. Finally, in Section 6, we shall outline our conclusions.

## 2. 'CLASSICAL' DISCRIMINANT ANALYSIS

In Discriminant Analysis $K$ categories $\Pi_1, \ldots, \Pi_K$, $K \geq 2$ are considered. Each category $\Pi_g$ is associated with a probability density $f_g(\mathbf{x} \mid \theta_g)$ on $R^p$, where $p$ denotes the number of characteristics considered, and may have an incidence rate or prior probability $\pi_g$.

In order to assign an element $e$ with observation vector $\mathbf{x}$ to one of these $K$ categories we need a classification rule corresponding to a division of $R^p$ into disjoint regions $R_1, \ldots, R_K$ $(\bigcup R_g = R^p)$. The general rule is :

$$e \mapsto \Pi_g \text{ if } \mathbf{x} \in R_g, \quad g = 1, \ldots, K. \tag{2.1}$$

Thus, the objective is to construct these regions using the $K$ training samples of categorized data.

Usually, we are faced with questions like:
Which form do the p.d.f.'s have? Do the $K$ samples come from $K$ populations or from the $K$ components of a mixture? How well does the derived decision rule perform?

### 2.1. General Formulation of DA

In the 'classical' approach to discriminant analysis the $K$ p.d.f.'s are assumed to be *multivariate normal* densities. If the population covariance matrices can be considered equal, we apply *Linear Discriminant Analysis* (LDA), otherwise *Quadratic Discriminant Analysis* (QDA). In case of a $K$-component mixture problem we choose the *Bayes assignment rule*, otherwise the *Maximum Likelihood* (ML) *assignment rule*, which can be considered equivalent to the Bayes rule with equal incidence rates for the $K$ categories.

Using unbiased or ML estimates of the usually unknown parameters we proceed to derive estimated *classification functions*, $d_g(\mathbf{x} \mid \hat{\psi}_g)$, where $\hat{\psi}_g = (\hat{\mu}_g, \hat{\Sigma}_g, \hat{\pi}_g)$, $g = 1, \ldots, K$. Based on these $K$ classification functions we divide the sample space $S$ into $K$ estimated regions $\hat{R}_g$ as follows:

$$\hat{R}_g = \{\mathbf{x} \in S \mid d_g(\mathbf{x} \mid \hat{\psi}_g) > d_l(\mathbf{x} \mid \hat{\psi}_l), \text{ for all } l \neq g\}, \quad g, l = 1, \ldots, K. \tag{2.2}$$

These regions $\hat{R}_g$ are separated by $(p-1)$-dimensional surfaces.

A separating surface between two adjacent regions $\hat{R}_g$ and $\hat{R}_l$ satisfies the equation

$$d_g(\mathbf{x} \mid \hat{\psi}_g) = d_l(\mathbf{x} \mid \hat{\psi}_l). \tag{2.3}$$

Then, $e \mapsto \Pi_g$

$$\text{if } d_g(\mathbf{x}|\hat{\psi}_g) = \max_l d_l(\mathbf{x}|\hat{\psi}_l), \text{ for all } l = 1, \ldots, K \ ,$$
$$\text{or if } 0 \leq d_g(\mathbf{x}|\hat{\psi}_g) - d_l(\mathbf{x}|\hat{\psi}_l), \text{ for all } l = 1, \ldots, K \ , \tag{2.4}$$

for $g = 1, \ldots, K$.
Using the $K$ training samples we evaluate the performance of the assignment rule classically by computing either an estimate of the *apparent error rate* based on the *resubstitution method*,

$$\hat{\eta}_{app} = \sum_{g=1}^{K} \hat{\pi}_g \sum_{\substack{l=1 \\ l \neq g}}^{K} \frac{m_{lg}}{n_g} \quad , \tag{2.5}$$

where $m_{lg}$ of the $n_g$ elements from sample $g$ lie in $\hat{R}_l$, or an estimate of the error rate based on the *leaving-one-out method* proposed by Lachenbruch and Mickey (1968),

$$\hat{\eta}_h = \sum_{g=1}^{K} \hat{\pi}_g \sum_{\substack{l=1 \\ l \neq g}}^{K} \frac{a_{lg}}{n_g} \quad , \tag{2.6}$$

where $a_{lg}$ is the number of hold-out elements from sample $g$ which lie in $\hat{R}_l$.

For more details on discriminant analysis, we refer the reader to Fahrmeir et al. (1984), Johnson and Wichern (1983), Lachenbruch (1975), Mardia et al. (1979), Seber (1984), and many other authors.

## 2.2. Problems inherent in DA

Although discriminant analysis as outlined above is well established in multivariate analysis, in practice, there are still deficiencies in the approaches to model selection and evaluation. These can be summarized under the following questions:

(i) Which parametric model fits best?

This question referring, for instance, to the assumed underlying structure of the group covariance matrices is usually answered by applying Box's M test (1949) using some arbitrarily set significance level. If the incidence rates of the $K$ categories are unknown, the decision concerning the assumption of equal or unequal incidence rates for the $K$ underlying categories is traditionally made in favour of the latter one.

(ii) Is discrimination worthwhile?

If all $K$ true mean vectors can be considered equal, discriminant analysis is pointless. This problem can be addressed using Wilks' $\Lambda$ criterion (1932) as the test statistic. Again, the level of significance is arbitrary, and, moreover, this test only applies to a model assuming equal covariance matrices.

(iii) What is the dimension of the discriminant space?

The general practice is to employ a variation of Wilks' $\Lambda$ to test for the significance of the discriminant functions without giving any clear indication of which groups are well separated. Furthermore, of course, the deficiencies mentioned under (ii) are present.

(iv) How is the performance of the assignment rule?

In the 'classical' approach the model is only evaluated with respect to potential misclassifications, that is, without considering the effect of actually misclassifying individual elements. Such an evaluation does not provide for a 'learning' system.

Clearly, there are many problems related to the 'classical' approach to discriminant analysis. Therefore, in Section 3 and 4, we shall introduce a general methodology called *Two-Stage Multi-Sample Cluster Analysis* (TSMSCA) as a general approach to DA. The best model is selected without making arbitrary assumptions and decisions. The only assumption on the data still made is that they come from a multivariate normal distribution. This can be tested using the multivariate measures of *skewness* and *kurtosis* [Mardia et al. (1979)]. If the data fail the tests, they can be transformed to near-normality, e.g. by using appropriate *power transformations* [Box and Cox (1964)].

## 3. THE GENERAL PROCEDURE OF TSMSCA

DA can be approached from a cluster analytic point of view. Searching for the appropriate dimension of the discriminant space can be viewed as finding the 'true' number of clusters in clustering samples of the categories under consideration. Such a point of view also emphasizes the relative differences among the categories, which in turn is the problem of *Multiple Comparison Procedures* (MCP). The possibility of using *Multi-Sample Cluster Analysis* (MSCA) as an alternative to MCPs was originally suggested by Bozdogan (1986). Classification of individuals, and, thus, the entire evaluation procedure can be viewed as clustering individuals. This approach evaluates the initially formed classification rules with respect to within-group homogeneity. In the literature, such an approach is discussed by Seber (1984, p.328), and by Ganesalingam and McLachlan (1979).

Our method is different and new as it represents a combined procedure of model selection and evaluation in an expert fashion. Therefore, we shall propose TSMSCA as a *general* approach to DA.

### 3.1. The Two-Stage Multi-Sample Cluster Problem

The problem of TSMSCA arises when we have several random samples from independently existing categories, and we intend to cluster these samples, and to investigate, or even improve upon within-cluster homogeneity.

Suppose each individual element, whose origin with respect to the $K$ categories is known, has been measured on $p$ characteristics. Let

$$\mathbf{X}(n \times p) = \begin{bmatrix} \mathbf{X}_1 \\ \mathbf{X}_2 \\ \vdots \\ \mathbf{X}_K \end{bmatrix} \begin{matrix} (n_1 \times p) \\ (n_2 \times p) \\ \vdots \\ (n_K \times p) \end{matrix} \qquad (3.1)$$

be the data matrix comprising all $K$ random samples, where $\mathbf{X}_g(n_g \times p)$ represents the observations from the $g$th group or sample, $g = 1, \ldots, K$, and $n = \sum_{g=1}^{K} n_g$. The objective of TSMSCA obviously consists of two stagewise connected goals. First, we search for $k$ homogeneous clusters of the $K$ samples for $k = 1, \ldots, K$. We try to obtain the smallest number $k^*$ in order to reduce the dimensionality of the data set, and, hence, the risk inherent in estimating the parameters of the model.

Then, considering each of the $k^*$ sets of $k$ clusters, $k = 1, \ldots, k^*$, as an initial partition, $\mathcal{C} = \{C_1, \ldots, C_k\}$, of the $n$ elements we reallocate them such that the homogeneity of the $k$ clusters is improved, or we just test their cluster membership with respect to within-cluster homogeneity.

In the first stage, depending on the cardinality of $K$, we generate all possible clustering alternatives, or use a splitting algorithm, to find the best clustering of samples. In the second stage, we employ an efficient transfer and swapping algorithm to reassign individual elements.

### 3.2. Grouping Samples

Let $K$ be the number of samples or groups, and $k$ be the number of non-empty clusters of samples, where $k \leq K$. Then, the *total number of ways of clustering K samples into k-sample clusters*, the order of $k$-sample clusters being irrelevant, is given by

$$\omega(K,k) = S(K,k) = \frac{1}{k!}\sum_{g=0}^{k}(-1)^g \left[ \begin{array}{c} k \\ g \end{array} \right] (k-g)^K \quad , \qquad (3.2)$$

which is known as the *Stirling Number of the Second Kind.* Since in MSCA $k^*$, the 'true' number of clusters of samples, or, in terms of DA, the number of distinct regions, is not known in advance, the total number of clustering alternatives becomes

$$\sum_{k=1}^{K} S(K,k). \qquad (3.3)$$

For detailed explanations and proofs, see, e.g., Duran and Odell (1974), and Späth (1975). For more on the representation and patterns of clustering alternatives in MSCA, see Bozdogan (1986).

On the basis of AIC and CAIC we can scrutinize the total number of possible clustering alternatives and select the ***best clustering alternatives*** for $k = 1, \ldots, K$-sample clusters. Moreover, among these best $k$-clustering alternatives we can detect the overall best alternative, thus, also yielding the best number of clusters or regions, $k^*$. Organizing these pieces of information into a *decision-tree classifier* we obtain a hierarchy of separation and assignment rules, which we shall discuss in more detail in Section 4.

If we do not want to generate all possible clustering alternatives, especially for rather big $K$, we directly construct the decision-tree classifier using the fact that $\omega(K,2) = 2^{K-1} - 1$.

The following *Splitting Algorithm* allows for moving from one optimal step to the next:

STEP–1 Start with $k = 1$-sample cluster, that is, when all $K$ samples are together in one cluster and compute AIC or CAIC.

STEP–2 Split the $k$=1-sample cluster $\omega(K,2)$ times into $k$=2-sample clusters, where $\omega(K,2)$ is the total number of ways of clustering $K$ samples into 2-sample clusters, and compute the AICs or CAICs of the resulting clustering alternatives. Choose the best $k=2$-clustering alternative by the minimum AIC or CAIC.

STEP–3 In order to find the best $k$-sample alternative clusters for $k=3,\ldots,K$, split each cluster $C$ of the best $(k-1)$-sample alternative clusters $\omega_c(m_c,2)=2^{m_c-1}-1$ times into 2-sample clusters, where $m_c$ is the number of samples in cluster $C$, while leaving the other $(k-2)$ clusters unchanged. Thus, generate $\sum_{c=1}^{k-1}\omega_c(m_c,2)$ $k$-sample alternative clusters and compute their AICs or CAICs. Then, choose the best $k$-sample alternative clusters by the minimum AIC or CAIC.

STEP–4 Repeat STEP–3 until all the samples are clustered in their own singleton clusters, or stop if the next best split does not yield any improvement upon the current minimum AIC or CAIC.

### 3.3. Clustering Individual Elements

The problem of the second stage of TSMSCA is a general problem of cluster analysis. We are given an initial $k$-clusters partition with $\mathcal{C}=\{C_1,\ldots,C_k\}$, and we are asked to investigate whether this partition represents a good fit for the entire data set, that is, whether the clusters are homogeneous and well-enough separated. This leads to a search for poor fits, that is, individual elements which should be reassigned to a different cluster.

Except for $k=1$ all best $k$-sample clusters solutions of the decision-tree classifier of the first stage are considered initial partitions for the search for an improved $k$-clusters solution. Thus, on each step of the decision-tree we can detect which elements should be reassigned with respect to within-cluster homogeneity. While walking through the decision-tree, we actually evaluate the quality of the decision tree, that is, the underlying relationship of the $K$ samples. In terms of discriminant analysis we assess the performance of the best $k$-assignment rule which devides the $R^p$ into $\hat{R}_k$ regions, $k=2,\ldots,k^*$.

Using a *cross-validation method* we only tentatively reassign individual elements. The *conditional clustering method*, however, actually executes transfers and swaps of individual elements, where, with respect to the mixture model, the condition is given by a prescribed clustering. [The term *conditional mixture model* was introduced by Sclove (1977), (1983).] This latter method represents a 'learning' approach since taking into account the changing group structures can be viewed as an estimation *process* of the classification rules.

The *transfer and swapping algorithm* applied in the method of conditional clustering consists of two distinct phases. The first phase transfers individual elements (one at a time) from one cluster to another. The second phase swaps two individual elements between their respective clusters. The general goal is to minimize AIC or CAIC.

On each step $k$, starting from the best $k$-sample clustering, each algorithmically possible transfer is tested as to its potential of improving upon the current reference value of the criterion. If the transfer proves beneficial, it is executed provided no empty cluster would be created.
When the value of the criterion cannot be improved by any further transfer, each algorithmically possible swap of two individual elements from different clusters is tested and executed if it further minimizes the criterion.
When no further swaps are beneficial, the present $k$-clusters partition is considered terminal, and the best $(k+1)$-sample clusters solution is tested for beneficial transfers and swaps. Instead of moving on, though, the transferring phase could be re-entered and the swapping phase subsequently, and so on, until there is no more improvement upon the criterion value. (With respect to the computing time, however, re-entry appears to be prohibitive.)
For more detailed explanations of this algorithm, see Eisenblätter (1987), and also Banfield and Bassill (1977), and Späth (1983).

In constructing confusion tables for all best $k$-sample clusters solutions we can contrast the results of the two stages, and calculate error rates. Moreover, this sequence of confusion tables could reveal some interesting aspects of the process of separating samples, that is, in terms of DA, of the effect of adding the separating surfaces in decreasing importance.

## 4. MODEL-SELECTION AND -EVALUATION IN TSMSCA

### 4.1. Model-Selection Criteria

In order to obtain a system which sequentially makes a number of decisions in an expert fashion, we need an objective and 'informed' criterion. Akaike [(1973), (1974), (1977), (1981)] developed such a criterion for choosing an optimal and parsimonious model over a set of competing models $\{M_k; k = 1,\ldots,K\}$. It is called *Akaike's Information Criterion* (AIC), and is defined as follows:

$$AIC(k) = -2\ln L[\hat{\theta}(k)] + 2m(k) \ , \tag{4.1}$$

where $L[\hat{\theta}(k)]$ is the maximized likelihood function of the observation vectors, and $m(k)$ is the number of independent parameters under the model $M_k$.

Since the AIC statistic is essentially an estimator of the risk of a model under the maximum likelihood estimation, it is minimized to select a model $M_k$ among the $K$ alternative models. In order to penalize overparameterization more stringently than AIC does, model selection in TSMSCA can alternatively be based on an asymptotically consistent form of AIC, called CAIC [Bozdogan (1987)]. It is defined as follows:

$$CAIC(k) = -2\ln L[\hat{\theta}(k)] + m(k)\ln n \ . \tag{4.2}$$

Like AIC this criterion is entropy-based. Taking a Bayesian approach Akaike (1979), and Schwarz (1978) developed the same criterion, known as *BIC*.

We next present the AICs for multivariate normal models in classification procedures, and outline how they help solving the problems inherent in DA. For more detailed explanations, we refer the reader to Bozdogan [(1983), (1984)], Bozdogan and Sclove (1984), and Eisenblätter (1987). CAICs are derived and used in a similar fashion.

### 4.2. Discriminant Analysis and AIC

Throughout this section we shall suppose that we have samples of elements known to derive from $K$ independent categories. We denote the data from the samples by the independent matrices $\mathbf{X}_1,\ldots,\mathbf{X}_K$, where the rows of $\mathbf{X}_g(n_g \times p)$ are assumed to be independent and identically distributed (i.i.d.) $N_p(\mu_g, \Sigma_g)$, $g = 1,\ldots,K$. Moreover, we shall suppose that the individual categories have incidence rates $\pi_1,\ldots,\pi_K$, and $\sum_{g=1}^K \pi_g = 1$.
In terms of the vector of parameters $\psi = (\mu_1,\ldots,\mu_K,\Sigma_1,\ldots,\Sigma_K, \pi_1,\ldots,\pi_K)$ we shall consider the following basic models:

$$
\begin{array}{ll}
\text{(i)} & \psi = (\mu_1, \ldots, \mu_K, \Sigma_1, \ldots, \Sigma_K, \pi_1, \ldots, \pi_K) \ , \\
\text{(ii)} & \psi = (\mu_1, \ldots, \mu_K, \Sigma, \ldots, \Sigma, \pi_1, \ldots, \pi_K) \ , \\
\text{(iii)} & \psi = (\mu_1, \ldots, \mu_K, \Sigma_1, \ldots, \Sigma_K, \pi, \ldots, \pi) \ , \\
\text{(iv)} & \psi = (\mu_1, \ldots, \mu_K, \Sigma, \ldots, \Sigma, \pi, \ldots, \pi) \ , \\
\text{(v)} & \psi = (\mu, \ldots, \mu, \Sigma, \ldots, \Sigma, \pi, \ldots, \pi) \ .
\end{array}
$$

The collection of these models can be viewed as special cases of finite mixture models. For more on this, see Bozdogan (1983), Titterington et al. (1985), and Symons (1981).

4.2.1. Identification of the best fitting parametric model.
If we let $\{\mathbf{x}_{ig}; i = 1, \ldots, n_g\}$ be the observation vectors of a random sample from the $g$-th category, then, the likelihood function of the combined sample data can be written in the form

$$L(\mathbf{X} \mid \psi) = \prod_{g=1}^{K} \prod_{i=1}^{n_g} \pi_g N_p(\mathbf{x}_{ig} \mid \mu_g, \Sigma_g) \ , \tag{4.3}$$

and the log likelihood function is

$$l(\mathbf{X} \mid \psi) = \sum_{g=1}^{K} n_g \ln \pi_g + \sum_{g=1}^{K} \sum_{i=1}^{n_g} \ln N_p(\mathbf{x}_{ig} \mid \mu_g, \Sigma_g) \ . \tag{4.4}$$

The *maximum likelihood estimates* (MLEs) of the parameters of the models under consideration are

$$
\begin{aligned}
\hat{\mu}_g &= \bar{\mathbf{x}}_g \quad , \quad \hat{\mu} = \bar{\mathbf{x}} \ , \\
\hat{\Sigma}_g &= \frac{1}{n_g} \mathbf{W}_g = \frac{1}{n_g} \sum_{i=1}^{n_g} (\mathbf{x}_{ig} - \bar{\mathbf{x}}_g)(\mathbf{x}_{ig} - \bar{\mathbf{x}}_g)' \ , \\
\hat{\Sigma} &= \frac{1}{n} \mathbf{W} = \frac{1}{n} \sum_{g=1}^{K} \mathbf{W}_g \ , \\
\hat{\pi}_g &= \frac{n_g}{n}, \text{ and } \ \hat{\pi} = \frac{1}{K} \ ,
\end{aligned}
$$

for $g = 1, \ldots, K$.
Substituting the MLEs into the log likelihood functions and simplifying yields the AICs of the five parametric models as follows:

$$AIC(\{\mu_g, \Sigma_g, \pi_g\})$$
$$= np\ln(2\pi) + \sum_{g=1}^{K} n_g \ln|n_g^{-1}\mathbf{W}_g| + np - 2\sum_{g=1}^{K} n_g \ln n_g +$$
$$+ 2n\ln n + 2[Kp + Kp(p+1)/2 + (K-1)] \; , \quad (4.5)$$

$$AIC(\{\mu_g, \Sigma, \pi_g\})$$
$$= np\ln(2\pi) + n\ln|n^{-1}\mathbf{W}| + np - 2\sum_{g=1}^{K} n_g \ln n_g +$$
$$+ 2n\ln n + 2[Kp + p(p+1)/2 + (K-1)] \; , \quad (4.6)$$

$$AIC(\{\mu_g, \Sigma_g, \pi\})$$
$$= np\ln(2\pi) + \sum_{g=1}^{K} n_g \ln|n_g^{-1}\mathbf{W}_g| + np + 2n\ln K +$$
$$+ 2[Kp + Kp(p+1)/2 + 1] \; , \quad (4.7)$$

$$AIC(\{\mu_g, \Sigma, \pi\})$$
$$= np\ln(2\pi) + n\ln|n^{-1}\mathbf{W}| + np + 2n\ln K +$$
$$+ 2[Kp + p(p+1)/2 + 1] \; , \quad (4.8)$$

$$AIC(\{\mu, \Sigma, \pi\})$$
$$= np\ln(2\pi) + n\ln|n^{-1}\mathbf{T}| + np + 2[p + p(p+1)/2] \; ,$$
$$\text{where } \mathbf{T} = \sum_{i=1}^{n}(\mathbf{x}_i - \bar{\mathbf{x}})(\mathbf{x}_i - \bar{\mathbf{x}})'. \quad (4.9)$$

The best fitting parametric model is identified by the minimum AIC.
If model (v) is chosen, we have to conclude that the categories are not distinct and, thus, any further analysis would be useless.
Parametric models (i) and (ii) imply the Bayesian approach to DA, and models (iii) and (iv) the ML or Bayesian approach with equal incidence rates.
Moreover, parametric models (ii) and (iv) can be viewed in terms of LDA, and models (i) and (iii) in terms of QDA.

For more on the derivation of the AICs, including those for known incidence rates, and their formal relationship to DA rules, see Eisenblätter (1987). AICs as replacements for conventional tests in MANOVA are explained by Bozdogan [(1984), (1986)].

4.2.2. Determination of the dimension of the discriminant space. Suppose we rejected parametric model (v), that is, we can claim that there is a difference in the $K$ categories. If we let $r^*$ denote the dimension of the hyperspace spanned by $\mu_1, \ldots, \mu_K$, we can state $r^* > 0$, but $r^* \leq K - 1$.
We now proceed to determine $r^*$, the dimension of the discriminant space, by finding $k^*$, the best number of sample clusters. If $k^* = K$, that is, $r^* = K - 1$, we say the $\mu_g$'s span the $R^p$, otherwise we say the $\mu_g$'s lie in an $r^*$-dimensional hyperspace.
Other than determining the dimension of the model, we are also interested in deriving a *hierarchy of separation* of the underlying categories, that is, sorting the $r^*$ $(p-1)$-dimensional separating surfaces in decreasing importance.
Applying the splitting algorithm of Section 3.2. on the best fitting parametric model identified in Section 4.2.1. we can accomplish both goals.

In selecting the best $k$-sample alternative clusters for $k = 1, \ldots, K$ we derive the hierarchy of separation by choosing the best splits in decreasing importance regarding their contribution to overall homogeneity. The dimension of the space, $r^*$, is determined by comparing the best $k$-sample alternative clusters as we move from $k = 1$ to $k = K$.
On any step $k+1$ we can decide whether $k^* = k$

$$\begin{aligned} &\text{if} \quad AIC(k\,|\,\psi) < AIC(k+1\,|\,\psi) \ , \\ &\text{or if} \quad AIC(k\,|\,\psi) - AIC(k+1\,|\,\psi) < 0 \ , \end{aligned} \tag{4.10}$$

or for model (i), for instance,

$$\begin{aligned} \text{iff} \quad & [n_{c^*} \ln|n_{c^*}^{-1}\mathbf{W}_{c^*}| - 2n_{c^*} \ln n_{c^*}] - \\ & - [\sum_{j=1}^{2} n_{c_j^*} \ln|n_{c_j^*}^{-1}\mathbf{W}_{c_j^*}| - 2n_{c_j^*} \ln n_{c_j^*}] \\ & < p(p+3) + 2 \ , \end{aligned} \tag{4.11}$$

where $C^*$ denotes the sample cluster to be split on step $k+1$.

On each step $k+1$ the best split of the best $k$-sample alternative clusters is chosen to be the one which is most likely to improve upon $AIC(k|\psi)$. Since $m$, the number of independent parameters, is the same for all alternatives on step $k+1$, we actually select that split $S(c^*)$ which maximally reduces the heterogeneity of the best $k$-sample clusters partition. Hence, for model (i), for instance,

$$C^* \begin{matrix} \nearrow & C_1^* \\ \searrow & C_2^* \end{matrix} \quad \text{if } S(c^*) = \max_c S(c), c = 1, \ldots, k \ , \tag{4.12}$$

where

$$\begin{aligned} S(c) &= n_c \ln|n_c^{-1}\mathbf{W}_c| - 2n_c \ln n_c - \\ &- \max_j\{\sum_{l=1}^{2} n_{c_{l_j}} \ln|n_{c_{l_j}}^{-1}\mathbf{W}_{c_{l_j}}| - 2n_{c_{l_j}} \ln n_{c_{l_j}}\} \ , \\ & j = 1, \ldots, \omega_c \ , \end{aligned} \tag{4.13}$$

where $\omega_c$ is the number of ways to split sample cluster $C$ as defined in Section 3.2.

4.2.3. Assignment rules and their assessment.

On each step $k$, $k = 2, \ldots, k^*$, a new element $e$ with observation vector $\mathbf{x}$ is allocated according to the *AIC–assignment rule:*

$e \mapsto \Pi_q$

$$\begin{aligned} &\text{if } AIC(e \in C'_q \,|\, \psi, k) \\ &\quad = \min_j AIC(e \in C'_j \,|\, \psi, k), \text{ for all } j = 1, \ldots, k \ , \\ &\text{or if } AIC(e \in C'_q \,|\, \psi, k) - AIC(e \in C'_j \,|\, \psi, k) \le 0 \ , \\ &\quad \text{for all } j = 1, \ldots, k \ , \end{aligned} \tag{4.14}$$

for $q = 1, \ldots, k$, where $C' = C \cup \{e\}$. For model (i), for instance, this rule becomes

$e \mapsto \Pi_q$

$$\begin{aligned} \text{iff} \quad & (n_q+1)\ln|(n_q+1)^{-1}\mathbf{W}_q + n_q(n_q+1)^{-1}(\mathbf{x}-\bar{\mathbf{x}}_q)(\mathbf{x}-\bar{\mathbf{x}}_q)'| - \\ & -(n_j+1)\ln|(n_j+1)^{-1}\mathbf{W}_j + n_j(n_j+1)^{-1}(\mathbf{x}-\bar{\mathbf{x}}_j)(\mathbf{x}-\bar{\mathbf{x}}_j)'| \\ \le\ & 2[(n_q+1)\ln(n_q+1) - (n_j+1)\ln(n_j+1)] \ , \\ & \text{for all } j = 1, \ldots, k \ , \end{aligned} \tag{4.15}$$

for $q = 1, \ldots, k$.

If we decided to use the information acquired from the new

individual for any future assignment, we would then update $\hat{\psi}_q$.

The rule in (4.14) delineates the sample space $S$ into $k$ regions,

$$\hat{R}_q = \{\mathbf{x}_e \in S \,|\, AIC(e \in C_q' \,|\, \psi, k) \leq AIC(e \in C_j' \,|\, \psi, k),\\ \text{for all } j \neq q\},\ j, q = 1, \ldots, k. \tag{4.16}$$

In order to evaluate the rule in (4.14) we can choose between two methods, a cross-validation method and the method of conditional clustering.

In the *cross-validation* technique we consider one individual element $i$ of the combined sample unclassified, and tentatively assign it to that cluster for which AIC attains a minimum. If we suppose $i \in C_l$, we substitute $C_l' = C_l \setminus \{i\}$ for $C_l$ in the present partition $C$. Then, the AIC-assignment rule translates into

$$i \mapsto C_q \text{ if } AIC(i \in C_q' \,|\, \psi, k)\\ = \min_j AIC(i \in C_j' \,|\, \psi, k),\ j = 1, \ldots, k\ , \tag{4.17}$$

for $q = 1, \ldots, k$, where $C' = C \cup \{i\}$.

Doing this for all $n$ elements yields the following estimated error rate:

$$\hat{\eta}_{cv} = \sum_{j=1}^{k} \hat{\pi}_j \sum_{\substack{l=1 \\ l \neq j}}^{k} \frac{a_{jl}}{n_j}\ , \tag{4.18}$$

where $a_{jl}$ of the $n_j$ elements from $C_j$ are incorrectly classified to $C_l$.

Another way to assess the best $r$-dimensional discriminant alternative, that is, the best $k$-sample clustering alternative, is to execute all misclassifications, and score the resulting partition using AIC. If this value is less than the AIC of the best $k$-sample clusters partition, we can state that the 'new' partition displays a more homogeneous group structure. In fact, for large discrepancies we may say that the best $r$-dimensional alternative is inappropriate for DA.

In the *conditional clustering* technique we consider all $n$ elements from the training samples unclassified. In the *transfer phase* we reassign individual element $i$ according to the rule in (4.17). Since we execute transfers, we change the clusters and, thus, the assignment rule as we proceed in the transfer algorithm.

The resulting partition represents the initial partition for the *swapping phase.*
The specific rule for swapping two individual elements, $i \in C_l$ and $j \in C_m$ in the present partition $\mathcal{C}$, is given as follows:

$$\begin{array}{l} i \mapsto C_m \\ j \mapsto C_l \end{array} \text{ if } AIC(i \in C'_m; j \in C'_l \,|\, \psi, k) < AIC(i \in C_l; j \in C_m \,|\, \psi, k) \quad , m \neq l \ , \qquad (4.19)$$

where $C'_l = C_l \setminus \{i\} \cup \{j\}$, and $C'_m = C_m \setminus \{j\} \cup \{i\}$.
Again, since we execute swaps, for which the condition in (4.19) holds, we change the clusters and, thus, the assignment rule after each swap.
Comparing the partition produced by the conditional clustering method with the best $k$-sample clusters partition we can compute the following estimated error rate:

$$\hat{\eta}_{cc} = \sum_{j=1}^{k} \hat{\pi}_j \sum_{\substack{l=1 \\ l \neq j}}^{k} \frac{m_{jl}}{n_j} \quad , \qquad (4.20)$$

where $m_{jl}$ is the number of elements from $C_j$ incorrectly assigned to $C_l$.
Since the conditional clustering method is designed to search for a better fit with respect to within-cluster homogeneity and between-cluster heterogeneity, that is, represents a kind of 'learning' approach, the AIC of the 'new' partition is always less than the AIC of the best $k$-sample clusters partititon. Again, a substantial discrepancy may indicate either that the initial clusters are not well-defined, and/or that the set of $p$ predictor variables is inappropriate.

## 5. NUMERICAL EXAMPLES

In this section we shall give two numerical examples, and study Two-Stage Multi-Sample Cluster Analysis as a general approach to Discriminant Analysis.
Since both data sets failed the tests for multivariate normality, we transformed them to near-normality.

### 5.1. TSMSCA on Data from Diabetes Research

The data set published by Andrews and Herzberg (1985) is composed of 145 patients belonging to three clinical categories, namely *overt diabetic* ($\Pi_1$), *chemical diabetic* ($\Pi_2$), and *normal* ($\Pi_3$), measured on relative weight, fasting plasma glucose, glucose area, insulin area, and SSPG.
For this data set we have $p = 5$ characteristics, $n_1 = 33$, $n_2 = 36$, $n_3 = 76$, and total $n = 145$. Since AIC and CAIC produced the same results, we shall give the AICs only.

(i) Identification of the best fitting parametric model:
The AICs under the five parametric models are as follows:

$$AIC(\{\mu_g, \Sigma_g, \pi_g\}) = -4354.3574^*, \tag{5.1}$$
$$AIC(\{\mu_g, \Sigma, \pi_g\}) = -4226.2157 , \tag{5.2}$$
$$AIC(\{\mu_g, \Sigma_g, \pi\}) = -4333.9583 , \tag{5.3}$$
$$AIC(\{\mu_g, \Sigma, \pi\}) = -4205.8165 , \tag{5.4}$$
$$AIC(\{\mu, \Sigma, \pi\}) = -4172.7031 . \tag{5.5}$$

The minimum AIC occurs under the model for separate categories with different shapes and incidence rates in (5.1). Therefore, we shall proceed in the analysis using the model with parameter vector $\psi = (\mu_1, \mu_2, \mu_3, \Sigma_1, \Sigma_2, \Sigma_3, \pi_1, \pi_2, \pi_3)$.

(ii) Determination of the dimension of the discriminant space:
Since the model in (5.5) was not selected, we conclude that $r^* > 0$. In order to determine the dimension of the discriminant space, and to derive the hierarchy of separation of the categories, we cluster the $K = 3$ samples into $k = 1, 2$, and 3-sample clusters. As there is a total of five possible clustering alternatives only, we computed all AICs, and obtained the results given in Table I.

Looking at Table I we see that the minimum AIC occurs at clustering alternative 4, that is, the best clustering is the $k=2$-sample clusters partition $(1)(2,3) \equiv (\Pi_1)(\Pi_2, \Pi_3)$. This indicates that according to the sample data there seem to be only $k^* = 2$ diabetic categories. Thus, we claim that there is only one 4-dimensional separating surface. This surface separates the overt diabetic patients from the collection of chemical diabetic and normal patients. Table I also reveals that separating $\Pi_2$ from $\Pi_3$ yields the second best model. According to AIC we design the decision-tree classifier as shown in Figure 1.

(iii) Evaluation of the best discriminant model:
Since $k^* = 2$, or equivalenty $r^* = 1$, we proceed to assess the

Table I

MSCA of $K = 3$ diabetic categories

| Alternative | Clustering | k | AIC($\{\mu_g, \Sigma_g, \pi_g\}$) |
|---|---|---|---|
| 1 | (1,2,3) | 1 | -4172.7031 |
| 2 | (1,2) (3) | 2 | -4309.5664 |
| 3 | (1,3) (2) | 2 | -4144.4818 |
| 4 | (1) (2,3) | 2 | -4363.9294* |
| 5 | (1) (2) (3) | 3 | -4354.3574 |

Note: * Minimum AIC

Figure 1

Decision-tree classifier for $K = 3$ diabetic categories

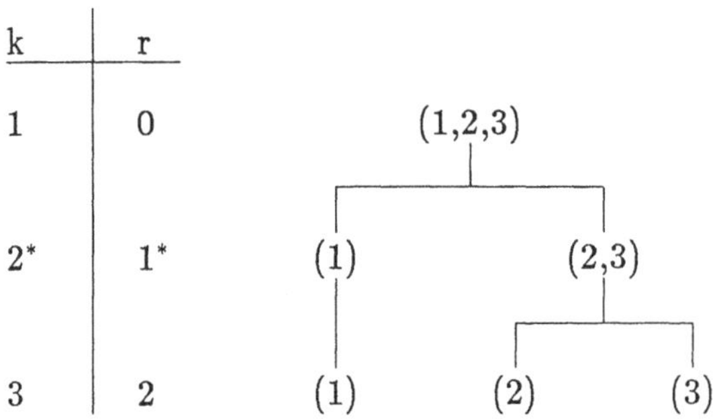

assignment rule based on alternative 4 in Table I. The assignment results of our two evaluation methods are summarized in Table II.

Based on the assignments in Table II we compute the evaluation statistics shown in Table III.

The information displayed in Table III tells us that even the application of conditional clustering yields a fairly low error rate for this data set. Although both AICs of the 'new' partitions have improved upon AIC$(k^* = 2)$, we may conclude that only the CC-method has produced clusters which are considerably more homogeneous than the $k^* = 2$-sample clusters. The CC result may ask for further analysis.

### 5.2. TSMSCA on Fisher's Iris Data

The iris data set published by Fisher (1936) consists of 150 irises belonging to three species, namely *Iris setosa* $(\Pi_1)$, *Iris versicolor* $(\Pi_2)$, and *Iris verginica* $(\Pi_3)$, measured on sepal and petal length and width.

For this data set we have $p = 4$ charcteristics, $n_1 = n_2 = n_3 = 50$, and total $n = 150$. Since AIC and CAIC selected different parametric models, we shall briefly present the results under both parametric models.

Table IV displays the best MSCA alternatives and their AICs and CAICs.

Since $k^* = 3$, we evaluate the assignment rules for $k = 2$, and $k = 3$. The evaluation statistics for the AIC-assignment rules are given in Table V, and those for the CAIC-assignment rules in Table VI.

The error rates for $k = 2$ indicate that Iris setosa is well-separated from the other two species. This is also true for $k = 3$ because the misclassified observations neither come from this category nor are they assigned to it. For $k^* = 3$ we can state that the error rates as well as the criteria values basically support the assignment rules based on the $k$-sample clusters, $k = 2, 3$.

## 6. CONCLUSIONS

From the numerical results in Section 5, we see that the minimum AIC and CAIC procedures can successfully select and evaluate the best assignment rules in Discriminant Analysis.

With this new approach it is now possible to determine *a priori* which parametric model should be used in the analysis of

Table II

Combined confusion table of cross-validation (CV) and conditional clustering (CC)

| Sample clusters | Assignments by the method of cross-validation $C_1'$ | $C_2'$ | conditional clustering $C_1'$ | $C_2'$ |
|---|---|---|---|---|
| $C_1$ (1) | 31 | 2 | 29 | 4 |
| $C_2$ (2) | 2 | 34 | 1 | 35 |
| (3) | 0 | 76 | 0 | 76 |
| Total | 33 | 112 | 30 | 115 |

Table III

Summary of evaluation statistics

| Method of evaluation | Number of misclassified individuals | Error rate (in %) | AIC($\{\mu_g, \Sigma_g, \pi_g\}$) |
|---|---|---|---|
| CV | 4 | 2.76 | -4377.6112 |
| CC | 5 | 3.45 | -4404.6986 |

Table IV

Best $k$-sample clusters alternatives for $K = 3$ Iris species

| k | Clustering | AIC($\{\mu_g, \Sigma_g, \pi\}$) | CAIC($\{\mu_g, \Sigma, \pi\}$) |
|---|---|---|---|
| 1 | (1,2,3) | -78.40274 | -36.25384 |
| 2 | (1) (2,3) | -214.12636 | -67.68664 |
| 3 | (1) (2) (3) | -367.26301* | -238.96057* |

Note: * Minimum AIC, or CAIC

Table V

Evaluation statistics for the AIC-assignment rules

| Method of evaluation | k | Number of misclassified individuals | Error rate (in %) | AIC($\{\mu_g, \Sigma_g, \pi\}$) |
|---|---|---|---|---|
| CV | 2 | 0 | 0.00 | -214.12636 |
| | 3 | 4 | 2.67 | -375.53261 |
| CC | 2 | 0 | 0.00 | -214.12636 |
| | 3 | 5 | 3.33 | -378.83153 |

Table VI

Evaluation statistics for the CAIC-assignment rules

| Method of evaluation | k | Number of misclassified individuals | Error rate (in %) | CAIC($\{\mu_g, \Sigma, \pi\}$) |
|---|---|---|---|---|
| CV & CC | 2 | 0 | 0.00 | -67.68664 |
| | 3 | 3 | 2.00 | -248.64638 |

a data set. We can identify the dimension of the discriminant space, and, thus, we can reduce the dimensionality of data sets as shown with the diabetic categories data set. In clustering samples into homogeneous sets of samples we can measure the amount of homogeneity and heterogeneity present in the discriminant space. All this can be accomplished without having to assume any arbitrary level of significance.

In our evaluation procedure we can measure the effect of assignments on the entire model. Moreover, our assignment rules can be easily updated for the addition of any new elements. Using conditional clustering we can even obtain partitions which better fit the data with respect to homogeneity.

In concluding, the new approach presented in this paper will provide the researcher with a concise, and more refined way of studying discrimination of samples and classification of individual elements for a particular multi-sample data set.

Therefore, we recommend the use of model-selection criteria in conjunction with Two-Stage Multi-Sample Cluster Analysis (TSMSCA) as a general approach to 'classical' discriminant analysis (DA).

Dorothea Eisenblätter
Seminar für Wirtschafts- und Sozialstatistik
der Universität zu Köln
Albertus-Magnus-Platz
5000 Köln 41
Federal Republic of Germany

Hamparsum Bozdogan
Department of Mathematics
Math./ Astronomy Building
University of Virginia
Charlottesville, Virginia 22903
U.S.A.

## REFERENCES

Akaike, H. (1973). 'Information Theory and an Extension of the Maximum Likelihood Principle' in *Second International Symposium on Information Theory*, (B.N. Petrov and F. Csaki, editors). Akademiai Kiado: Budapest, 267–281.

Akaike, H. (1974). 'A New Look at the Statistical Model Identification,' *IEEE Transactions on Automatic Control* **AC-19**, 716–723.

Akaike, H. (1977). 'On Entropy Maximization Principle' in *Proceedings on Applications of Statistics* (P.R. Krishnaiah, editor). North-Holland: Amsterdam, 27–47.

Akaike, H. (1979). 'A Bayesian Analysis of the Minimum AIC Procedure,' *Annals of the Institute of Statistical Mathematics (Part A)* **30**, 9–14.

Akaike, H. (1981). 'Likelihood of a Model and Information Criteria,' *Journal of Econometrics* **16**, 3–14.

Andrews, D.F., and Herzberg, A.M. (1985). *Data. A Collection of Problems from Many Fields for the Student and Research Worker*. Springer: New York.

Banfield, C.F., and Bassill, L.C. (1977). 'Algorithm AS 113: A Transfer Algorithm for Non-hierarchical Classification,' *Applied Statistics* **26**, 206–210.

Box, G.E.P. (1949). 'A General Distribution Theory for a Class of Likelihood Criteria,' *Biometrika* **36**, 317–346.

Box, G.E.P., and Cox, D.R. (1964). 'An Analysis of Transformations,' (with discussion), *Journal of the Royal Statistical Society (B)* **26**, 211–252.

Bozdogan, H. (1983). 'Determining the Number of Component Clusters in the Standard Multivariate Normal Mixture Model Using Model-Selection Criteria,' Technical Report No. UIC/DQM/A83-1, June 16, 1983, Army Research Office Contract DAAG29-82-K-0155, University of Illinois at Chicago, Box 4348, Chicago, Illinois 60680.

Bozdogan, H. (1984). 'AIC-Replacements for Multivariate Multi-Sample Conventional Tests of Homogeneity Models,' Technical Paper #4 in Statistics, Department of Mathematics, University of Virginia, Charlottesville, VA, 22903.

Bozdogan, H. (1986). 'Multi-Sample Cluster Analysis as a General Alternative to Multiple Comparison Procedures,' *Bulletin of Informatics and Cybernetics Research Association of Statistical Sciences* **22**, 95–130.

Bozdogan, H. (1987). 'Model Selection and Akaike's Information Criterion (AIC): The General Theory and Its Analytical Extensions,' (to appear in the Special Issue of *Psychometrika*).

Bozdogan, H., and Sclove, S.L. (1984). 'Multi-Sample Cluster Analysis Using Akaike's Information Criterion,' *Annals of the Institute of Statistical Mathematics (Part B)* **36**, 243–253.

Duran, B.S., and Odell, P.L. (1974). *Cluster Analysis: A Survey.* Springer: New York.

Eisenblätter, D. (1987). *Two-Stage Multi-Sample Cluster Analysis*, Ph.D. Thesis (anticipated), Seminar für Wirtschafts- und Sozialstatistik der Universität zu Köln.

Fahrmeir, L., and Hamerle, A., editors (1984). *Multivariate statistische Verfahren.* de Gruyter: Berlin.

Fisher, R.A. (1936). 'The Use of Multiple Measurements in Taxonomic Problems,' *Annals of Eugenics* **7**, 179–188.

Ganesalingam, S., and McLachlan, G.J. (1979). 'A Case Study of Two Clustering Methods Based on Maximum Likelihood,' *Statistical Neerlandica* **33**, 81–90.

Johnson, R.A., and Wichern, D. (1983). *Applied Multivariate Statistical Analysis.* Prentice Hall: New York.

Lachenbruch, P.A. (1975). *Discriminant Analysis.* Hafner Press: New York.

Lachenbruch, P.A., and Mickey, M.R. (1968). 'Estimation of Error Rates in Discriminant Analysis,' *Technometrics* **10**, 1–11.

Mardia, K.V., Kent, J.T., and Bibby, J.M. (1979). *Multivariate Analysis.* Academic Press: New York.

Schwarz, G. (1978). 'Estimating the Dimension of a Model,' *Annals of Statistics* **6**, 461–464.

Sclove, S.C. (1977). 'Population Mixture Models and Clustering Algorithms,' *Communications in Statistics A* **6**, 417–434.

Sclove, S.C. (1983). 'Application of the Conditional Population Mixture Model to Image Segmentation,' *IEEE Transactions on Pattern Analysis and Machine Intelligence* **PAMI-5**, 428–433.

Seber, G.A. (1984). *Multivariate Observations.* Wiley: New York.

Späth, H. (1975). *Cluster-Analyse-Algorithmen.* Oldenbourg: München.

Späth, H. (1983). *Cluster-Formation und -Analyse.* Oldenbourg: München.

Symons, M.J. (1981).. 'Clustering Criteria and Multivariate Normal Mixtures,' *Biometrics* **37**, 35–43.

Titterington, D.M., Smith, A.F.M., and Makov, U.E. (1985). *Statistical Analysis of Finite Mixture Distributions.* Wiley: New York.

Wilks, S.S. (1932). 'Certain Generalization in the Analysis of Variance,' *Biometrika* **24**, 471–494.

Yasunori Fujikoshi

# ON RELATIONSHIP BETWEEN THE AIC AND THE OVERALL ERROR RATES FOR SELECTION OF VARIABLES IN A DISCRIMINANT ANALYSIS

## SUMMARY

This paper deals with the problem of selecting the "best" subset of variables in a discriminant analysis with the aim of allocating future observations, in the context of two multivariate normal populations with the same covariance matrix. We consider the methods based on the following three criteria: (i) the AIC for the "no additional information" model, (ii) the overall error rate criterion based on the linear classification statistic and (iii) the overall error rate criterion based on the ML classification statistic. It is shown that there is a close relationship between the AIC and the overall error rate criteria.

## 1. INTRODUCTION

Various methods have been proposed in a discriminant analysis whose aim is the description of differences among the populations concerned or the allocation of future observations. For a summary, see example, R. J. McKay and N. A. Campbell (1982a, b). In this paper we deal with the problem of selecting the "best" subset of variables in a discriminant analysis with the aim of allocating future observations, in the context of two multivariate normal populations with the same covariance matrix.

Suppose that individuals belong to one of two populations, $\Pi_1$ and $\Pi_2$, and that $\underset{\sim}{x} = (x_1,\ldots,x_p)'$ represents a full set of p random variables. We assume that in $\Pi_g$, $\underset{\sim}{x}$ is distributed as $N_p[\underset{\sim}{\mu}^{(g)}, \Sigma]$, where $\underset{\sim}{\mu}^{(1)}$, $\underset{\sim}{\mu}^{(2)}$ and $\Sigma$ are unknown. Suppose that random samples of size $N_g$ from each population $\Pi_g$ (g = 1, 2) are available. Let $\bar{\underset{\sim}{x}}^{(1)}$, $\bar{\underset{\sim}{x}}^{(2)}$ and S be the sample means and the unbiased pooled sample covariance matrix based on the samples,



*H. Bozdogan and A. K. Gupta (eds.), Multivariate Statistical Modeling and Data Analysis, 121–138.*

$$X = [\underset{\sim}{x}_1^{(1)}, \ldots, \underset{\sim}{x}_{N1}^{(1)}, \underset{\sim}{x}_1^{(2)}, \ldots, \underset{\sim}{x}_{N2}^{(2)}]. \tag{1.1}$$

Our problem is to select the "best" subset of variables for classifying an observation $\underset{\sim}{x}$ as coming from $\Pi_1$ or $\Pi_2$. For a subset $j = \{j_1,\ldots,j_{k(j)}\}$ of the set of subscripts 1, 2, ..., p, we define

$$\underset{\sim}{x}(j) = (x_{j_1}, \ldots, x_{j_{k(j)}})'. \tag{1.2}$$

We will identify j by $\underset{\sim}{x}(j)$. Let J be the family of all possible subsets of $\{1,\ldots,p\}$. Then our problem may be stated as the one of selecting an optimum subjet j from a subfamily of J or J itself.

In this paper we consider the methods based on the following three criteria: (i) the AIC (Akaike's information criterion (Akaike, 1973)) for the "no additional information" model, (ii) the overall error rate criterion based on the linear classification statistic and (iii) the overall error rate criterion based on the ML classification statistic. In Section 2 we give the derivation of the AIC. In Section 3 we obtain the new criteria which are asymptotic estimates for the overall error rates with general a priori probabilities. It may be noted that McLachlan (1980) has proposed an asymptotic estimate for the overall error rate with equal a priori probabilites associated with the linear classification statistic. Fujikoshi (1985a, b) has shown that there is a close relationship between the methods based on the AIC and the overall error rate criterion. In Section 4 it is also shown that there is a close relationship between the AIC and the overall error rate criteria with general a priori probabilities, by deriving asymptotic distributions of the criteria. We point out some difference between the overall error rate criteria associated with the linear classification statistic and the ML classification statistic.

We will use the following notations. We denote the $\underset{\sim}{\mu}^{(g)}$, $\bar{\underset{\sim}{x}}^{(g)}$, $\Sigma$ and S specified by j or $\underset{\sim}{x}(j)$ by $\underset{\sim}{\mu}^{(g)}(j)$, $\bar{\underset{\sim}{x}}^{(g)}(j)$, $\Sigma(j)$ and S(j), respectively. Let $\Delta$ and $\Delta(j)$ be the population Mahalanobis distance between $\Pi_1$ and $\Pi_2$ based on $\underset{\sim}{x}$ and $\underset{\sim}{x}(j)$, respectively. Further, let D and D(j) be the sample quantities corresponding to $\Delta$ and $\Delta(j)$, respectively.

## 2. THE AIC FOR SELECTION OF VARIABLES

We shall denote the pdf of the observation matrix X in (1.1) by $f(X;\ \Theta)$. As a natural approach for our problem, it is possible to proceed the following steps:

(i) to construct a family of models H(j) which is closely relating to selection of variables and is expressed in terms of $\Theta$.

(ii) to apply a criterion for model choice, which choose the best model in a predictive sense, to the family of models.

It is natural to define H(j) by the "no additional information" model which implies that $\underset{\sim}{x}(j)$ has as much information about our statistical analysis as $\underset{\sim}{x}$ as follows:

$$H(j)\ :\ \Delta = \Delta(j) \tag{2.1}$$

As a criterion for model choice suitable for the step (ii), we use Akaike's information criterion. Under the approriate conditions, the criterion is equivalent to the following: choose the model H(j) to minimize

$$\mathrm{AIC}(j) = -2\log\{f(X;\ \hat{\theta}_j)\} + 2p(j), \tag{2.2}$$

where $\hat{\Theta}_j$ is the maximum likelihood estimate of $\Theta$ under H(j), and p(j) is the number of independent parameters under H(j).

It is known (Fujikoshi (1983)) that for the family of the models H(j) in (2.1),

$$\begin{aligned} A(j) &= \mathrm{AIC}(j) - \mathrm{AIC}(\{1,\ldots,p\}) \\ &= N\log\{\ 1 + (p-k(j))F(j)/(N-p-1)\} \\ &\quad - 2(p-k(j)), \end{aligned} \tag{2.3}$$

where $N = N_1 + N_2$, $n = N - 2$ and

$$F(j) = \frac{N-p-1}{p-k(j)} \cdot \frac{D^2 - D(j)^2}{n(N_1^{-1}+N_2^{-1}) + D(j)^2}\ . \tag{2.4}$$

The first term in A(j) is $-2\log\{$the likelihood ratio criterion for H(j)$\}$, whose asymptotic null distribution is a chi-square distribution with (p-k(j)) degrees of freedom. The likelihood ratio test is equivalent to use the F(j) in (2.4), which is distributed according to an F distribution with degrees of freedom p-k(j) and N-p-1, respectively, under H(j).

First we outline the derivation (Fujikoshi (1983)) of (2.3) by a direct application of (2.2). For simplicity, we consider the case of $j = \bar{k} = \{1,\ldots,k\}$. Consider the partitions $\underset{\sim}{x} = (\underset{\sim}{x}_1', \underset{\sim}{x}_2')$, $\underset{\sim}{x}_1 = (x_1,\ldots,x_k)'$, and

$$\underset{\sim}{\mu}^{(g)} = \begin{pmatrix} \underset{\sim}{\mu}_1^{(g)} \\ \underset{\sim}{\mu}_2^{(g)} \end{pmatrix}, \quad \Sigma = \begin{pmatrix} \Sigma_{11} & \Sigma_{12} \\ \Sigma_{21} & \Sigma_{22} \end{pmatrix} \tag{2.5}$$

conforming with the partition of $\underset{\sim}{x}$. It is well known (Rao (1962)) that $H(\bar{k})$ is equivalent to

$$\underset{\sim}{\mu}_2^{(1)} - \beta\underset{\sim}{\mu}_1^{(1)} = \underset{\sim}{\mu}_1^{(2)} - \beta\underset{\sim}{\mu}_1^{(2)} \quad (= \underset{\sim}{\mu}_{2\cdot 1}) \tag{2.6}$$

where $\beta = \Sigma_{21}\Sigma_{11}^{-1}$. Considering the conditional density of $\underset{\sim}{x}_2$ given $\underset{\sim}{x}_1$, we have

$$\begin{aligned} -2\log\{f(X;\,\Theta)\} &= Np\log(2\pi) + \\ &+ N\log|\Sigma_{11}| + \textstyle\sum_{g=1}^{2}\sum_{r=1}^{N_g} \mathrm{tr}\,\Sigma_{11}^{-1}(\underset{\sim}{x}_{1r}^{(g)} - \underset{\sim}{\mu}^{(g)})(\underset{\sim}{x}_{1r}^{(g)} - \underset{\sim}{\mu}^{(g)})' + \\ &+ N\log|\Sigma_{22\cdot 1}| + \textstyle\sum_{g=1}^{2}\sum_{r=1}^{N_g} \mathrm{tr}\,\Sigma_{22\cdot 1}^{-1}(\underset{\sim}{x}_{2r}^{(g)} - \underset{\sim}{\mu}_{2\cdot 1} - \beta\underset{\sim}{x}_{1r}^{(g)}) \\ &\qquad \times (\underset{\sim}{x}_{2r}^{(g)} - \underset{\sim}{\mu}_{2\cdot 1} - \beta\underset{\sim}{x}_{1r}^{(g)})', \end{aligned} \tag{2.7}$$

where $\Sigma_{22\cdot 1} = \Sigma_{22} - \Sigma_{21}\Sigma_{11}^{-1}\Sigma_{12}$. Let

$$\begin{aligned} W &= \textstyle\sum_{g=1}^{2}\sum_{r=1}^{N_g}(\underset{\sim}{x}_r^{(g)} - \bar{\underset{\sim}{x}}^{(g)})(\underset{\sim}{x}_r^{(g)} - \bar{\underset{\sim}{x}}^{(g)})', \\ T &= \textstyle\sum_{g=1}^{r}\sum_{r=1}^{N_g}(\underset{\sim}{x}_r^{(g)} - \bar{\underset{\sim}{x}})(\underset{\sim}{x}_r^{(g)} - \bar{\underset{\sim}{x}})', \end{aligned} \tag{2.8}$$

where $\bar{\underset{\sim}{x}} = N^{-1}(N_1\bar{\underset{\sim}{x}}^{(1)}+N_2\bar{\underset{\sim}{x}}^{(g)})$. We partition W and T in the same way as in (2.5). Then we can write the MLE of $\Theta$ under $H(\bar{k})$ as follows:

$$\hat{\underset{\sim}{\mu}}_1^{(g)} = \bar{\underset{\sim}{x}}_1^{(g)}, \qquad N\hat{\Sigma}_{11} = W_{11}$$

$$\hat{\underset{\sim}{\mu}}_{2\cdot1} = \bar{\underset{\sim}{x}}_2 - \hat{\beta}\bar{\underset{\sim}{x}}_1, \qquad \hat{\beta} = T_{21}T_{11}^{-1} \tag{2.9}$$

$$N\hat{\Sigma}_{22\cdot1} = T_{22\cdot1} = T_{22} - T_{21}T_{11}^{-1}T_{12}.$$

The dimensionality of $\Theta$ under $H(\bar{k})$ is $p(\bar{k}) = (1/2)p(p+1) + 2k + (p-k)$. Hence we obtain

$$\begin{aligned} A(\bar{k}) &= N\log\{|W_{22\cdot1}|/|T_{22\cdot1}|\} - 2(p-k) \\ &= N\log\{[|W|/|T|]/[|W_{11}|/|T_{11}|]\} - 2(p-k). \end{aligned}$$

The final expression as in (2.3) is obtained by using

$$|T| = |W|\{1 + [(N_1N_2)/(Nn)]D^2\},$$

$$|T_{11}| = |W_{11}|\{1 + [(N_1N_2)/(Nn)]D(\bar{k})^2\}.$$

Next we derive (2.3), based on Akaike's idea, not by a direct use of (2.2). We can measure the goodness of the model H(j) by the predictive density for future observations Y or the log-density assessment. Here we assume that

$$Y = [\underset{\sim}{y}_1^{(1)}, \ldots, \underset{\sim}{y}_{N1}^{(1)}, \underset{\sim}{y}_1^{(2)}, \ldots, \underset{\sim}{y}_{N2}^{(2)}],$$

and Y has the same distribution as X and is independent of X. In a concrete form, we consider

$$R_A(j) = E_X E_Y[-2\log\{f(Y;\hat{\Theta}_j)\}] \tag{2.10}$$

as a measure for the goodness of the model H(j). The term 2p(j) in the AIC may be regarded as a correction term when we estimate $R_A(j)$ by $-2\log\{f(X;\hat{\Theta}_j)\}$. More precisely, the correction term is defined by

$$c(j) = E_X E_Y[-2\log\{f(Y;\hat{\Theta}_j)/f(X;\hat{\Theta}_j)\}] \qquad (2.11)$$

when H(j) is true. We evaluate c(j) in our problem. For simplicity, we consider $c(\bar{k})$. It is easy to see that

$$\begin{aligned} c(\bar{k}) &= E_X[N\,\mathrm{tr}\,\hat{\Sigma}^{-1}\Sigma + \\ &\quad + \textstyle\sum_{g=1}^{2} N_g\,\mathrm{tr}\,\hat{\Sigma}^{-1}(\hat{\mu}^{(g)}-\mu^{(g)})(\hat{\mu}^{(g)}-\mu^{(g)})'] - Np \\ &= E_X[N\,\mathrm{tr}\,\hat{\Sigma}_{11}^{-1}\Sigma_{11} + \\ &\quad + N\,\mathrm{tr}\,\hat{\Sigma}_{22\cdot1}^{-1}\{\Sigma_{22\cdot1}+(\hat{\beta}-\beta)\Sigma_{11}(\hat{\beta}-\beta)'\} + \\ &\quad + \textstyle\sum_{g=1}^{2} N_g\{\mathrm{tr}\,\hat{\Sigma}_{11}^{-1}(\hat{\mu}_1^{(g)}-\mu_1)(\hat{\mu}_1^{(g)}-\mu_1)' + \\ &\quad + \mathrm{tr}\,\hat{\Sigma}_{22\cdot1}^{-1}[\hat{\mu}_{2\cdot1}-\mu_{2\cdot1}+(\hat{\beta}-\beta)\mu_1^{(g)}][\hat{\mu}_{2\cdot1}-\mu_{2\cdot1}+ \\ &\qquad\qquad + (\hat{\beta}-\beta)\mu_1^{(g)}]'\}] - Np, \end{aligned}$$

where the MLE's of $\Theta$ are defined by (2.9). We use the following Lemma.

Lemma 2.1. Under the model $H(\bar{k})$ we have the following results:

(i) $W_{11}$ is distributed according to $W_k(\Sigma_{11}, n)$,

(ii) $T_{22\cdot1}$ is distributed according to $W_{p-k}(\Sigma_{22\cdot1}, N-k-1)$,

(iii) The elements of $U_1 = \Sigma_{22\cdot1}^{-1/2}[\hat{\mu}_{2\cdot1}-\mu_{2\cdot1}, \hat{\beta}-\beta](A'A)^{1/2}$ are independently distributed according to N(0, 1), where $A = [\mathbf{1}, X_1]$, $\mathbf{1} = (1,\ldots,1)'$ and $X = [X_1', X_2']'$, $X_1: k \times N$,

(iv) The elements of $U_2 = \Sigma_{22\cdot1}^{-1/2}(\hat{\beta}-\beta)T_{11}^{1/2}$ are independently distributed according to N(0, 1),

(v) $T_{22\cdot1}$, $U_1$(or $U_2$), $T_{11}$ and $\bar{x}_1$ are independent.

Proof. (i) is well known. The conditional distribution of $X_2'$ given $X_1'$ is

$$N_{N(p-k)}\left(A\begin{pmatrix}\underset{\sim}{\mu}_{2\cdot 1}' \\ \beta'\end{pmatrix},\quad I_N \otimes \Sigma_{22\cdot 1}\right).$$

Further we have

$$X_2(I_N - A(A'A)^{-1}A')X_2' = T_{22\cdot 1},$$

$$X_2A(A'A)^{-1} = [\hat{\underset{\sim}{\mu}}_{2\cdot 1}, \hat{\beta}].$$

By using these properties and the ordinary technique as in MANOVA, we can obtain (ii) $\sim$ (v).

Using Lemma 3.1, we have

$$\begin{aligned} c(\bar{k}) = N[ & \frac{k(N+2)}{N-k-1} + \frac{(p-k)(N+1)}{N-p-2} - p + \\ & + \frac{p-k}{N-p-2}\{\mathrm{tr}\, \tilde{T}_{11}^{-1}[(N+1)I_k + \frac{N_1N_2}{N}\tilde{\underset{\sim}{\delta}}_1\tilde{\underset{\sim}{\delta}}_1']\}], \end{aligned} \tag{2.12}$$

where $\tilde{T}_{11} = \Sigma_{11}^{-1/2}T_{11}\Sigma_{11}^{-1/2}$ and $\tilde{\underset{\sim}{\delta}}_1 = \Sigma_{11}^{-1/2}(\underset{\sim}{\mu}_1^{(1)} - \underset{\sim}{\mu}_1^{(2)})$. This result implies the following:

(i) As $N_1 \to \infty$, $N_2 \to \infty$ and $N_1/N_2 \to$ a positive limit,

$$\begin{aligned} c(j) &= p(p+1) + 2p + 2k(j) + O(N^{-1}) \\ &= 2p(j) + O(N^{-1}), \end{aligned}$$

(ii) Under the additional assumption of $\tilde{\underset{\sim}{\delta}}_1 = \underset{\sim}{0}$,

$$\begin{aligned} c(j) = [ & 2p(j)N^2 - \{(2k(j)+5)p^2 + \\ & + (8k(j)+15)p + k(j)(k(j)+5)\}N + \\ & + (k(j)+2)(K(j)+3)p^2 + 2(2k(j)^2+8k(j)+9)p + \end{aligned}$$

$$+ 2k(j)(k(j)+1)]/ \{(N-p-2) \times (N-k(j)-2)(N-k(j)-3)\}. \tag{2.13}$$

When N is not large, we can correct the A(j) in (2.3) as

$$\tilde{A}(j) = N \log\{1 + (p-k(j))F(j)/(N-p-1)\} + + c(j) - c(\{1, \ldots, p\}), \tag{2.14}$$

where c(j) is given by (2.13).

## 3. THE OVERALL ERROR RATE CRITERIA

For the selection of variables with the aim of allocating new observations, it is natural to select the subset of variables minimizing the overall error rate in a classification procedure. When the classification procedure is based on the linear classification statistic and a subst of variables $\underset{\sim}{x}(j)$, then we may classify a new observation $\underset{\sim}{x}$ by means of the classification statistic

$$w(j) = (\bar{\underset{\sim}{x}}^{(1)}(j) - \bar{\underset{\sim}{x}}^{(2)}(j))'S(j)^{-1}\{\underset{\sim}{x}(j) - - \frac{1}{2}(\bar{\underset{\sim}{x}}^{(1)}(j) + \bar{\underset{\sim}{x}}^{(2)}(j))\}. \tag{3.1}$$

The rule is to classify $\underset{\sim}{x}$ as coming from $\Pi_1$ if $w(j) > 0$ and from $\Pi_2$ if $w(j) \leq 0$. Then we have the actual overall error rate with a priori probabilities q and 1-q

$$L_{w(q)}(j; X) = q\, P(w(j) \geq 0 \mid \underset{\sim}{x} \in \Pi_1, X) + + (1-q)P(w(j) < 0 \mid \underset{\sim}{x} \in \Pi_2, X).$$

We may consider

$$R_{w(q)}(j) = E_X[L_{w(q)}(j; X)] \tag{3.2}$$

as a natural risk when $\underset{\sim}{x}(j)$ is used.

Another classification statistic concerned is the ML classification statistic

$$\begin{aligned} z(j) = & \frac{N_1}{N_1+1}(\underset{\sim}{x}(j)-\bar{\underset{\sim}{x}}^{(1)}(j))'S(j)^{-1}(\underset{\sim}{x}(j)-\bar{\underset{\sim}{x}}^{(1)}(j)) - \\ & - \frac{N_2}{N_2+1}(\underset{\sim}{x}(j)-\bar{\underset{\sim}{x}}^{(2)}(j))'S(j)^{-1}(\underset{\sim}{x}(j)-\bar{\underset{\sim}{x}}^{(2)}(j)). \end{aligned} \tag{3.3}$$

The rule is to classify $\underset{\sim}{x}$ as coming from $\Pi_1$ if $z(j) \leq 0$ and from $\Pi_2$ if $z(j) \geq 0$. The risk corresponding to (3.2) is

$$R_{z(q)}(j) = E_X[L_{z(q)}(j;\ X)], \tag{3.4}$$

where

$$\begin{aligned} L_{z(q)}(j) = & \ q\,P(z(j) \geq 0 \mid \underset{\sim}{x} \in \Pi_1,\ X) + \\ & + (1-q)P(z(j) < 0 \mid \underset{\sim}{x} \in \Pi_2,\ X). \end{aligned}$$

In most cases, the a priori probabilies are not known, and in practice, we use

$$\text{(i) } q = 1/2 \quad \text{or} \quad \text{(ii) } q = N_1/N.$$

Okamoto (1963) and Memon and Okamoto (1971) obtained asymptotic expansions of the distributions of $w(j)$ and $z(j)$ when $N_1 \to \infty$, $N_2 \to \infty$, and $N_1/N_2 \to \rho$ (a positive constant). Using these results we obtain the following expansions for $R_{w(q)}(j)$ and $R_{z(q)}(j)$:

$$\begin{aligned} R_{w(q)}(j) = & \ \Phi(-\tfrac{1}{2}\Delta(j)) + \\ & + \phi(-\tfrac{1}{2}\Delta(j))[\frac{1}{16N_1}\{\Delta(j) + 4(4q-1)\frac{k(j)-1}{\Delta(j)}\} + \\ & + \frac{1}{16N_2}\{\Delta(j) + 4(3-4q)\frac{k(j)-1}{\Delta(j)}\} + \\ & + \frac{1}{4n}(k(j)-1)\Delta(j)] + O(N^{-2}), \end{aligned} \tag{3.5}$$

$$R_{z(q)}(j) = \Phi(-\frac{1}{2}\Delta(j)) + $$
$$+ \phi(-\frac{1}{2}\Delta(j))[\frac{1}{16N_1}\{(3-4q)\Delta(j) + \frac{4(k(j)-1)}{\Delta(j)}\} + $$
$$+ \frac{1}{16N_2}\{(4q-1)\Delta(j) + \frac{4(k(j)-1}{\Delta(j)}\} + \tag{3.6}$$
$$+ \frac{1}{4n}(k(j)-1)\Delta(j)] + O(N^{-2}),$$

where $\Phi(\cdot)$ and $\phi(\cdot)$ are the cumulative distribution function and the probability density function of the standard normal distribution. When $q = 1/2$, we have

$$R_{w(1/2)}(j) = R_{z(1/2)}(j) + O(N^{-2}).$$

However, when $q \neq 1/2$, the two risks are different in the term of $O(N^{-1})$. When $H(j)$ is true, it holds that

$$R_{w(q)}(\{1,\ldots,p\}) - R_{w(q)}(j)$$
$$= \frac{(p-k(j))}{4\Delta}\phi(-\frac{1}{2}\Delta)[(4q-1)\frac{1}{N_1} + \tag{3.7}$$
$$+ (3-4q)\frac{1}{N_2} + \Delta^2\frac{1}{n}] + O(N^{-2}),$$

$$R_{z(q)}(\{1,\ldots,p\}) - R_{z(q)}(j)$$
$$= \frac{(p-k(j))}{4\Delta}\phi(-\frac{1}{2}\Delta)[\frac{1}{N_1} + \frac{1}{N_2} + \tag{3.8}$$
$$+ \Delta^2\frac{1}{n}] + O(N^{-2}).$$

In general, it is preferable that the left-hand side of (3.7) or (3.8) is positive, so that we can select smaller subsets of variables. It may be noted that neglecting the term of $O(N^{-2})$, the right-hand side of (3.8) is always positive, and the one of (3.7) is positive in most cases, but in some case can be negative.

Now we obtain asymptotic unbiased estimators for $R_{w(q)}(j)$ and $R_{z(q)}(j)$. It is well known (McLachlan (1973)) that

$$\begin{aligned} E[\Phi(-\tfrac{1}{2}D(j))] &= \Phi(-\tfrac{1}{2}\Delta(j)) + \\ &+ \phi(-\tfrac{1}{2}\Delta(j))[\frac{\Delta(j)}{32n}\{\Delta(j)^2 - 4(2k(j)+1)\} + \\ &+ \frac{1}{16\Delta(j)}(\frac{1}{N_1}+\frac{1}{N_2})\{\Delta(j)^2 - 4(k(j)-1)\}] + \\ &+ O(N^{-2}). \end{aligned} \tag{3.9}$$

Using (3.7) $\sim$ (3.9) and noting $E[D(j)] = \Delta(j) + O(N^{-1})$, we can get the following asymptotic unbiased estimators $M_{w(q)}(j)$ and $M_{z(q)}(j)$:

$$M_{w(q)}(j) = \Phi(G_{w(q)}(j)), \tag{3.10}$$

$$M_{z(q)}(j) = \Phi(G_{z(q)}(j)), \tag{3.11}$$

where

$$\begin{aligned} G_{w(q)}(j) &= -\tfrac{1}{2}D(j) + \frac{k(j)-1}{D(j)}(\frac{q}{N_1}+\frac{1-q}{N_2}) + \\ &+ \frac{D(j)}{32n}\{4(4k(j)-1) - D(j)^2\}, \end{aligned} \tag{3.12}$$

$$\begin{aligned} G_{z(q)}(j) &= -\tfrac{1}{2}D(j) + \frac{k(j)-1}{D(j)}(\frac{1}{2N_1}+\frac{1}{2N_2}) + \\ &+ \frac{D(j)}{8}(\frac{1-2q}{N_1}+\frac{2q-1}{N_2}) + \\ &+ \frac{D(j)}{32n}\{4(4k(j)-1) - D(j)^2\}. \end{aligned} \tag{3.13}$$

These estimators satisfy

$$E[M_{w(q)}(j)] = R_{w(q)}(j) + O(N^{-2}),$$

$$E[M_{z(q)}(j)] = R_{z(q)}(j) + O(N^{-2}).$$

The estimator $M_{w(1/2)}(j)$ was proposed by McLachlan (1980). We note that

$$M_{w(1/2)}(j) = M_{z(1/2)}(j).$$

We denote the selection methods based on $M_{w(q)}(j)$ and $M_{z(q)}(j)$ by $\hat{j}_{w(q)}$ and $\hat{j}_{z(q)}$ respectively, i.e.,

$$M_{w(q)}(\hat{j}_{w(q)}) = \mathrm{Min}\ M_{w(q)}(j),$$

$$M_{z(q)}(\hat{j}_{z(q)}) = \mathrm{Min}\ M_{z(q)}(j).$$

Since $\Phi$ is a monotone increasing function, $\hat{j}_{w(q)}$ and $\hat{j}_{z(q)}$ minimize also $G_{w(q)}(j)$ and $G_{z(q)}(j)$, respectively.

## 4. ASYMPTOTIC DISTRIBUTIONS OF THE THREE CRITERIA

We consider the selection methods based on the three criteria $A(j)$, $M_{w(q)}(j)$ and $M_{z(q)}(j)$ and denote these selections by $\hat{j}_A$, $\hat{j}_{w(q)}$ and $\hat{j}_{z(q)}$, respectively. Fujikoshi (1985) showed that there is a close relationship between $\hat{j}_A$ and $\hat{j}_{w(1/2)}$, by obtaining their asymptotic distributions. A selection method $\hat{j}_*$ may be regarded as a mapping from the observation matrix X to a subfamily of J.

We treat the case when the subfamily concerned is

$$\bar{J} = \{\bar{1}, \bar{2}, \ldots, \bar{p}\},$$

where $\bar{k} = \{1,\ldots,k\}$. This is the case when an assesment of the relative importance of individual variables is given a priori and the initial order of the variables $x_1,\ldots,x_p$ make sense. For selection method $\hat{j}_*$, we are interested in

$$p_{N,*}(\bar{m}) = P(\hat{j}_* = \bar{m}), \quad \bar{m} = \bar{1}, \bar{2}, \ldots, \bar{p}. \tag{4.1}$$

We may assume, without loss of generality,

$$\begin{aligned} &\text{Assumption:}\ \ H(\bar{k}) \text{ is true, i.e., } \Delta(\bar{k}) = \Delta \\ &\qquad \text{and } \Delta(\bar{m}) < \Delta \text{ for } \bar{m} = \bar{1}, \ldots, \overline{k-1}. \end{aligned} \tag{4.2}$$

Under this Assumption we consider the asymptotic distribution of $\hat{j}_*$ in the case

$$N_1 \to \infty,\ N_2 \to \infty, \text{ and } N_1/N_2 \to \rho \text{ (a positive limit).} \tag{4.3}$$

Since $D(j) \to \Delta(j)$ in probability, it is easy to see

$$\lim P(A(\bar{i}) \leq A(\bar{m})) = 0$$

for any $\bar{i}$ and $\bar{m}$ such that $1 \leq i < k \leq m \leq p$. This property holds also for $M_{W(q)}(j)$ and $M_{Z(q)}(j)$. Hence we have the following Theorem.

Theorem 4.1. Suppose the Assumption in (4.2) is satisfied. Then, it holds that for each $\hat{j}_*$ of the selection methods $\hat{j}_A$, $\hat{j}_{W(q)}$ and $\hat{j}_{Z(q)}$,

(i) $\lim p_{N,*}(\bar{m}) = 0, \quad \bar{m} = \bar{1}, \ldots, \overline{k-1},$

(ii) $\lim p_{N,*}(\bar{m}) = \lim \tilde{p}_{N,*}(\bar{m}), \quad \bar{m} = \bar{k}, \ldots, \bar{p},$

where $\tilde{p}_{N,*}(\bar{m}) = P(\tilde{j}_* = m)$ and $\tilde{j}_*$ means the selection in the case when $\bar{J}$ is restricted to $\tilde{J} = \{\bar{k}, \overline{k+1}, \ldots, \bar{p}\}$.

Now we consider the asymptotic distributions of $\tilde{j}_A$, $\tilde{j}_{W(q)}$ and $\tilde{j}_{Z(q)}$ which depend only on the statistics

$$\{D(\bar{k}), D(\overline{k+1}), \ldots, D(\bar{p})\}. \tag{4.3}$$

We use the following Lemma.

Lemma 4.1. Suppose that the model $H(\bar{k})$ is true. Then, without loss of generality, we may assume

$$\Sigma = I_p, \quad \underset{\sim}{\mu}^{(1)} - \underset{\sim}{\mu}^{(2)} = (\Delta, 0, \ldots, 0)' = \underset{\sim}{\delta} \tag{4.4}$$

when we treat the distributions of the statistics based on $\{D(\bar{k}), D(\overline{k+1}), \ldots, D(\bar{p})\}$.

Proof. Let $B_{22}$ be a lower triangular matrix such that $B_{22}\Sigma_{22\cdot1}B_{22}' = I_{p-k}$. Let $B_{11} = H\Sigma_{11}^{-1/2}$, and $B_{21} = -B_{22}\Sigma_{21}\Sigma_{11}^{-1}$, where $H$ is an orthogonal matrix such that $B_{11}(\underset{\sim}{\mu}_1^{(1)} - \underset{\sim}{\mu}_1^{(2)}) = [\Delta, 0, \ldots, 0]$. Then, letting

$$B = \begin{bmatrix} B_{11} & 0 \\ B_{21} & B_{22} \end{bmatrix},$$

we can see that

$$B(\underset{\sim}{\mu}^{(1)} - \underset{\sim}{\mu}^{(2)}) = \underset{\sim}{\delta}, \quad B\Sigma B' = I_p .$$

The Lemma is proved by noting that the statistics $\{D(\bar{k}),\ldots, D(\bar{p})\}$ are invariant under the transformation,

$$\bar{\underset{\sim}{x}}^{(g)} \to B\bar{\underset{\sim}{x}}^{(g)} + \underset{\sim}{b}, \quad S \to BSB' .$$

From Lemma 4.1 we may assume (4.4) when we treat the distributions of $\tilde{j}_A$, $\tilde{j}_{W(q)}$ and $\tilde{j}_{Z(q)}$. In the following we shall do that. Let

$$\begin{aligned} \underset{\sim}{d} &= \bar{\underset{\sim}{x}}^{(1)} - \bar{\underset{\sim}{x}}^{(2)} = \underset{\sim}{\delta} + \frac{1}{\sqrt{n}}\underset{\sim}{y} , \\ S &= I_p + \frac{1}{\sqrt{n}}V . \end{aligned} \tag{4.5}$$

Then, we have the following:

(i) $\underset{\sim}{y} = (y_1,\ldots,y_p)'$ is distributed according to $N_p[\underset{\sim}{0}, n(\frac{1}{N_1} + \frac{1}{N_2})I_p]$,

(ii) The limiting distribution of $V = [v_{\ell m}]$ is normal with mean 0 and $E[v_{\ell\ell}^2] = 2$, $E[v_{\ell m}^2] = 1$, $\ell \neq m$.

(iii) $\underset{\sim}{y}$ and $V$ are independent.

From (4.5) we can write

$$\begin{aligned} D(\bar{m}) &= [(\underset{\sim}{\delta}(\bar{m})+\frac{1}{\sqrt{n}}\underset{\sim}{y}(\bar{m}))'(I_m+\frac{1}{\sqrt{n}}V(\bar{m}))^{-1}(\underset{\sim}{\delta}(\bar{m})+\frac{1}{\sqrt{n}}\underset{\sim}{y}(\bar{m}))]^{1/2} \\ &= \Delta + \frac{1}{2\Delta\sqrt{n}}(2\Delta y_1 - \Delta^2 v_{11}) + \frac{1}{2\Delta n}\textstyle\sum_{\ell=1}^{m}(\Delta v_{1\ell}-y_\ell)^2 \\ &\quad - \frac{1}{8\Delta^3 n}(2\Delta y_1-\Delta^2 v_{11})^2 + O_p(N^{-3/2}). \end{aligned} \tag{4.6}$$

Let

$$z_\ell = (\Delta v_{1\ell} - y_\ell)/\sigma_N, \qquad \ell = 1,\ldots,p, \tag{4.7}$$

where $\sigma_N = \sigma_N(1/2)$ and

$$\sigma_N(q) = [n\{\frac{2q}{N_1} + \frac{2(1-q)}{N_2}\} + \Delta^2]^{1/2}.$$

Then, using (4.6), we obtain the following expressions for any $i > m \geq k$:

$$\begin{aligned} n\{&G_{w(q)}(\bar{m}) - G_{w(q)}(\bar{i})\} \\ &= \frac{\sigma_N^2}{4\Delta}\{\textstyle\sum_{\ell=m+1}^{i} z_\ell^2 - 2(i-m)(\sigma_N(q)/\sigma_N)^2\} + O_p(N^{-1}), \end{aligned} \tag{4.8}$$

$$\begin{aligned} n\{&G_{z(q)}(\bar{m}) - G_{z(q)}(\bar{i})\} \\ &= \frac{\sigma_N^2}{4\Delta}\{\textstyle\sum_{\ell=m+1}^{i} z_\ell^2 - 2(i-m)\} + O_p(N^{-2}), \end{aligned} \tag{4.9}$$

$$\begin{aligned} A(&\bar{m}) - A(\bar{i}) \\ &= \textstyle\sum_{\ell=m+1}^{i} z_\ell^2 - 2(i-m) + O_p(N^{-2}). \end{aligned} \tag{4.10}$$

Based on these results, we can express our asymptotic results in terms of

$$\begin{aligned} s_\alpha(m) &= P(\cap_{\ell=1}^{m}(U_\ell > 0)), \\ t_\alpha(m) &= P(\cap_{\ell=1}^{m}(U_\ell \leq 0)), \end{aligned} \tag{4.11}$$

where $s_\alpha(0) = t_\alpha(0) = 1$, $U_\ell = (V_1 - 2\alpha) + \cdots + (V_\ell - 2\alpha)$ and $V_i$'s are independent random variables with $\chi_1^2$ distributions. For further reductions of $s_\alpha(m)$ and $t_\alpha(m)$, see Spitzer (1956) and Shibata (1976).

Theorem 4.2. Under the Assumption in (4.2) it holds that for $\bar{m} = \bar{k}, \overline{k+1}, \ldots, \bar{p}$,

$$\lim p_{N,A}(\bar{m}) = s_1(m-k)t_1(p-m),$$
$$\lim p_{N,w(q)}(\bar{m}) = s_\alpha(m-k)t_\alpha(p-m), \qquad (4.12)$$
$$\lim p_{N,z(q)}(\bar{m}) = s_1(m-k)t_1(p-m),$$

where $\alpha = 1 + \beta$ and

$$\beta = \{\rho(1-2q) + \rho^{-1}(2q-1)\}/\{2 + \rho + \rho^{-1} + \Delta^2\}$$

Proof. We shall derive only the case of $\hat{j}_{W(q)}$ since we can derive the other cases similarly. From Theorem 4.1. (ii) and the monotonicity of $\Phi(\cdot)$ we have

$$\begin{aligned} &\lim P(\hat{j}_{w(q)} = \bar{m}) \\ &= \lim P(\tilde{j}_{w(q)} = \bar{m}) \\ &= \lim P(M_{w(q)}(\bar{m}) \leq M_{w(q)}(\bar{i});\ \bar{i} = \bar{k},\ldots,\bar{p}) \\ &= \lim P(G_{w(q)}(\bar{m}) \leq G_{w(q)}(\bar{i});\ \bar{i} = \bar{k},\ldots,\bar{p}). \end{aligned}$$

We can write the last expression as

$$\begin{aligned} &\lim P(\{[G_{w(q)}(\bar{i}) - G_{w(q)}(\bar{m})] \geq 0;\ \bar{i} = \bar{k},\ldots,\overline{m-1}\} \\ &\quad \text{and } \{[G_{w(q)}(\bar{m}) - G_{w(q)}(\bar{i})] \leq 0;\ \bar{i} = \overline{m+1},\ldots,\bar{p}\}), \end{aligned}$$

and hence, using (4.8) we obtain the final result.

It may be noted that the asymptotic distribution of $\hat{j}_{W(q)}$ depends on a priori probabilities, but the asymptotic distribution of $\hat{j}_{Z(q)}$ does not depend on a priori probabilites, and is the same as the one of $\hat{j}_A$. When $q = 1/2$, the two selections $\hat{j}_{W(q)}$ and $\hat{j}_{Z(q)}$ are identical, and have the same asymptotic distribution as $\hat{j}_A$. When $q = N_1/N$, $\alpha \leq 1$ and hence

$$\begin{aligned} \lim P(\hat{j}_A = \bar{k}) &= \lim P(\hat{j}_{z(q)} = \bar{k}) \\ &\geq \lim P(\hat{j}_{w(q)} = \bar{k}). \end{aligned}$$

It is interesting to point out that each of the three selections has the property,

$$\lim P(\hat{j}_* = \bar{m}) = 0, \quad \bar{m} = \bar{1},\ldots,\overline{k-1},$$

$$\lim P(\hat{j}_* = \bar{k}) \neq 1,$$

$$\lim P(\hat{j}_* = \bar{m}) > 0, \quad \bar{m} = \overline{k+1},\ldots,\bar{p}.$$

We note that there is also a similar relationship between the three selections in the case when $\bar{J}$ is any subfaimily or J itself. For the relationship between $\hat{j}_{W(1/2)}$ and $\hat{j}_A$, see Fujikoshi (1985 a).

Department of Mathematics
Faculty of Science
Hiroshima University
Hiroshima 730, Japan

## REFERENCES

Akaike, H. (1973). 'Information theory and an extension of the maximum likelihood principle'. In: 2nd International Symposium on Information Theory (B. N. Petrov and F. Czáki, eds.), pp.267-281, Akademiai Kiadó, Budapest.

Fujikoshi, Y. (1983). 'A criterion for variable selection in multiple discriminant analysis'. Hiroshima Math. J. 13, 203-214.

Fujikoshi, Y. (1985 a). 'Selection of variables in two-group discriminant analysis by error rate and Akaike's information criteria'. J. Multiv. Anal. 17, 27-37.

Fujikoshi, Y. (1985 b). 'Selection of variables in discriminant analysis and canonical correlation analysis. In: Multivariate Analysis - VI (P. R. Krishnaiah, ed.), pp.219-236, North-Holland.

McLachlan, G. J. (1973). 'An asymptotic expansion of the expectation of the estimated error rate in discriminant analysis'. Austral. J. Statist. 15, 210-214.

McLachlan, G. J. (1980). 'On the relationship between the F test and the overall error rate for variable selection in two-group discriminant analysis'. Biometrics 36, 501-510.

McKay, R. J. and Campbell, N. A. (1982). 'Variable selection techniques in discriminant analysis I. Description'. British J. Math. Statist. Psychology 35, 1-29.

Mckay, R. J. and Campbell, N. A. (1982). 'Variable selection techniques in discriminant analysis II. Allocation'. British J. Math. Statist. Psychology 35, 30-41.

Memon, A. Z. and Okamoto, M. (1971). 'Asymptotic expansion of the distribution of the Z statistic in discriminant analysis'. J. Multiv. Anal. 1, 294-307.

Okamoto, M. (1963). 'An asymptotic expansion for the distribution of the linear discriminant function'. Ann. Math. Statist. 34, 1286-1301.

Rao, C. R. (1970). 'Inference on discriminant function coefficients'. In: Essays in Prob. and Statist. (R. C. Bose, ed.), pp.587-602. Univ. of North Carolina Press, Chapel Hill.

Shibata, R. (1976). 'Selection of the order of an autoregressive model by Akaike's information criterion'. Biometrika 63, 117-126.

Spitzer, F. (1956). A combinatorial lemma and its application to probability theory. Trans. Amer. Math. Soc. 82, 323-339.

A. K. Gupta and J. Tang

# DISTRIBUTION OF LIKELIHOOD CRITERIA AND BOX APPROXIMATION

## ABSTRACT

In this paper, the exact distribution of a random variable whose moments are a certain function of gamma functions (Box, 1949), has been derived. It is shown that Box's asymptotic expansion can be obtained from this exact distribution by collecting terms of the same order. From the point of view of computation, the derived series has a distinct advantage over the results of Box since the coefficients satisfy a recurrence relation.

Key words and phrases: Asymptotic Distribution, Percentage Points, Convergence, Recursive Formula, Test of Independence.

## 1. INTRODUCTION

Consider a statistic $W$ $(0 \leq W \leq 1)$ whose $h^{th}$ moment is given by

$$E\,W^h = K \left[ \frac{\prod_{j=1}^{b} y_j^{y_j}}{\prod_{k=1}^{a} x_k^{x_k}} \right]^h \frac{\prod_{k=1}^{a} \Gamma[x_k(1+h)+\xi_k]}{\prod_{j=1}^{b} \Gamma[y_j(1+h)+\eta_j]}, \quad h = 0,1,2,\ldots, \tag{1.1}$$

where $K$ is a constant such that $E\,W^0 = 1$, and

$$\sum_{k=1}^{a} x_k = \sum_{j=1}^{b} y_j .$$

Since $0 \leq W \leq 1$, the moments uniquely determine the distribution. Box (1949) derived an asymptotic expansion for the cumulative distribution function (c.d.f.) of $W$. The first three terms of the expansion are given by (23) of Anderson(1984, p. 315). Finding higher order terms of the expansion is not easy. Nevertheless



*H. Bozdogan and A. K. Gupta (eds.), Multivariate Statistical Modeling and Data Analysis, 139–145.*

this approximation is quite often used in multivariate statistical analysis. But this approximation is not accurate for small samples.

In section 2 of this paper we derive the general asymptotic expansion for the cumulative distribution function of W. In section 3 the percentage points obtained from this expansion are compared with some available exact percentage points.

## 2. DISTRIBUTION OF W

Here we follow the notation used by Anderson (1984, sec. 8.5). The following result will be needed in our derivation of the distribution of W.

**LEMMA 2.1.** *If*

$$g(x) = \sum_{k=1}^{m} \alpha_k x^{-k}$$

*then the series expansion for* $\exp[g(x)]$ *is given by*

$$\exp[g(x)] = \sum_{j=0}^{\infty} \beta_j x^{-j},$$

*where the coefficients* $\beta_j$*'s satisfy the following recurrence relation:*

$$\beta_0(m) = 1, \ \beta_j(m) = \frac{1}{j} \sum_{k=1}^{\min(j,m)} k\alpha_k \beta_{j-k}(m) \qquad j = 1,2,\dots . \tag{2.1}$$

*Proof:* Differentiate both sides of (2.1) with respect to $x$ and compare the coefficients of $x^{-(j+1)}$.

Now let $M = -2 \log W$, then the characteristic function of $\rho M$ $(0 < \rho < 1)$ is given by (see Anderson 1984, p. 313)

$$\phi(t) = (1 - 2it)^{-\frac{1}{2}f} \exp[\sum_{r=1}^{m} \omega_r (1 - 2it)^{-r} - \sum_{r=1}^{m} \omega_r + R_{m+1}], \tag{2.2}$$

where

$$f = -2[\sum_{k=1}^{a} \xi_k - \sum_{j=1}^{b} \eta_j - \frac{1}{2}(a - b)]; \ i = \sqrt{-1};$$

$$\omega_r = \frac{(-1)^{r+1}}{r(r+1)} [\sum_{k=1}^{a} \frac{B_{r+1}(\beta_k + \xi_k)}{(\rho x_k)^r} - \sum_{j=1}^{b} \frac{B_{r+1}(\varepsilon_j + \eta_j)}{(\rho y_j)^r}], \ r = 1,2,\dots,m,$$

$\varepsilon_j = (1 - \rho)y_j$, $\beta_k = (1 - \rho)x_k$, $j = 1,2,\ldots,b$, $k = 1,2,\ldots,a$; $B_r(h)$ is the Bernoulli polynomial of degree $r$ and order unity defined by

$$\frac{\tau e^{h\tau}}{e^{\tau} - 1} = \sum_{r=0}^{\infty} \frac{\tau^r}{r!} B_r(h) \text{ for } |\tau| < 2\pi$$

and

$$R'_{m+1} = \sum_{r=0}^{a} 0(x_k^{-(m+1)}) + \sum_{j=1}^{b} 0(y_j^{-(m+1)}).$$

Using Lemma 2.1, we get

$$\exp[\sum_{r=1}^{m} \omega_r (1 - 2it)^{-r}] = \sum_{j=0}^{\infty} u_j(m)(1 - 2it)^{-j},$$

and

$$\exp[-\sum_{r=1}^{m} \omega_r] = (\sum_{j=0}^{\infty} u_j(m))^{-1} = \sum_{j=0}^{\infty} v_j(m), \text{ say,}$$

where

$$u_0(m) = 1, \; u_j(m) = \frac{1}{j} \sum_{r=1}^{\min(m,j)} r\,\omega_r u_{j-r}(m); \; j = 1,2,\ldots, \tag{2.3}$$

and

$$v_0(m) = 1, \; v_j(m) = - \sum_{r=0}^{j-1} v_r(m) u_{j-r}(m); \; j = 1,2,\ldots. \tag{2.4}$$

Then (2.2) can be written as

$$\phi(t) = (1 - 2it)^{-\frac{1}{2}f} \{\sum_{j=0}^{\infty} u_j(m)(1 - 2it)^{-j}\}\{\sum_{j=0}^{\infty} v_j(m)\}$$

$$= (1 - 2it)^{-\frac{1}{2}f} \{\sum_{k=0}^{m} T_k(t) + R'''_{m+1}\} \tag{2.5}$$

where

$$T_k(t) = \sum_{r=0}^{k} u_r(m)v_{k-r}(m)(1 - 2it)^{-r}, \quad k = 0,1,2,\ldots,m$$

(see section 8.5, (17) and (18) of Anderson (1984)).

In most applications, we will have $x_k = c_k\theta$ and $y_j = d_j\theta$, where $c_k$ and $d_j$ will be constant and $\theta$ will vary (that is, will grow with sample size). In this case if $\rho$ is chosen so $(1 - \rho)x_k$ and $(1 - \rho)y_j$ have limits, then $R'''_{m+1}$ is $0(\theta^{-(m+1)})$ and $T_k(t) = 0(\theta^{-k})$.

Applying the Fourier inversion formula to (2.5), we obtain

$$\begin{aligned} P\{M \le M_0\} &= P\{\rho M \le \rho M_0\} \\ &= \sum_{k=0}^{m} \left(\sum_{r=0}^{k} u_r(m)v_{k-r}(m)\right)P\{\chi^2_{f+2r} \le \rho M_0\} + R^{v}_{m+1} \\ &= \sum_{r=0}^{m} \left(u_r(m) \sum_{j=0}^{m-r} v_j(m)\right)P\{\chi^2_{f+2r} \le \rho M_0\} + R^{v}_{m+1} \\ &= \sum_{r=0}^{m} \ell_r(m)P\{\chi^2_{f+2r} \le \rho M_0\} + R^{v}_{m+1}, \text{ say} \end{aligned} \tag{2.6}$$

where

$$\ell_r(m) = u_r(m) \sum_{j=0}^{m-r} v_j(m),$$

which can be computed recursively from (2.3)-(2.4), and the error

$$R^{v}_{m+1} = \int_0^{\rho M_0} \int_{-\infty}^{\infty} \frac{1}{2\pi}(1 - 2it)^{-\frac{1}{2}f} R'''_{m+1} e^{-itz} dt\, dz$$

is $0(\theta^{-(m+1)})$. The first three terms of (2.6) are given explicitly in (23), p. 315 of Anderson(1984). In many applications $\rho$ may be chosen such that $\omega_1 = 0$, although it does not matter what value $\rho$ takes as long as $0 < \rho \le 1$.

If we assume $x_k > 1$ and $y_j > 1$ and let $m \to \infty$ in (2.6), then it can be shown that the infinite series converge for $M_0 \in (0, 4\rho\beta\pi)$, where $\beta = \min_{k,j}\{x_k, y_j\}$; see Tang and Gupta (1987).

## 3. COMPARISON AND REMARKS

Suppose the random vector $X = (X_1, X_2, \ldots, X_p)'$ has a p-dimensional multivariate normal distribution. Let X be partitioned into q groups: group one containing $X_1, \ldots, X_{p_1}$, group two containing $X_{p_1+1}, \ldots, X_{p_2}$, etc. Let the matrix of sample correlation coefficients R, based on a sample of size N, be partitioned according to the partition of X; i.e., $R = (R_{ij})$, $i,j = 1,2,\ldots,q$, where $R_{ij}$ $(i \neq j)$ is the matrix of correlation coefficients between variables in the $i^{th}$ group and variables in the $j^{th}$ group, and $R_{ii}$ is the matrix of correlation coefficients of the $i^{th}$ group.

We are interested in testing independence of groups. Wilks (1932) derived the likelihood ratio test statistic for this hypothesis which is given by

$$\Lambda = |R| / \prod_{i=1}^{q} |R_{ii}|,$$

where $|A|$ denotes the determinant of the matrix A.

Let $n = N - 1$ and $p_i = i$ for $i = 1,2,\ldots,q$; then the moments of $\Lambda^{n/2}$ satisfy (1.1) with $a = b = p - 1$, $x_k = n/2$, $y_j = n/2$, $\zeta_k = -q/2$, $\eta_j = 0$. Applying the result of section 2 we obtain the c.d.f. of $\rho M = -\rho n \log \Lambda$ as follows:

$$P\{\rho M \leq v_0\} = \sum_{j=0}^{m} \ell_j(m) P\{\chi^2_{f+2j} \leq v_0\} + O(n^{-(m+1)}) \tag{3.1}$$

where $f = q(q-1)/2 = p(p-1)/2$. Then $\rho = 1 - (2q+5)/6n$ satisfies $\omega_1 = 0$.

Using the calculus of residues, Mathai and Katiyar (1979) obtained exact 5 and 1 percentage points of $U = -\{n - (2q+5)/6\} \log \Lambda$ for $q = 3(1)10$. Table I provides comparison between Mathai and Katiyar's exact values and the values obtained from (3.1) truncated at $m = 2$ and $m = 10$. They are denoted by MK, B(2), and B(10), respectively.

From Table I we see that if sufficiently many terms are included, the Box approximation gives exact percentage points in $(0, 4\rho\beta\pi)$. As q increases, the number of points outside this interval grows and hence the percentage points for small n fall outside the interval of convergence and become inaccurate. This provides a theoretical explanation of the remark made by Mudholkar, Trivedi and Lin (1982).

In case that $x_k$ and $y_j$ are positive as we have in this example, (1.1) can be transformed to Wilks' type-B integral equation for which the exact solution in two different expresions are available; see Gupta (1977), Walster and Tretter (1980), Tang and Gupta (1984, 1987), and Gupta and Tang (1984).

Table I

Comparison of Percentage Points of U for $q = 5$

| | 5% | | | 10% | | |
|---|---|---|---|---|---|---|
| n | B(2) | B(10) | MK | B(2) | B(10) | MK |
| 5 | 20.64* | 23.98* | 24.01 | 26.12* | 31.82* | 32.16 |
| 6 | 19.61 | 20.44 | 20.44 | 24.94* | 26.50* | 26.50 |
| 7 | 19.12 | 19.45 | 19.45 | 24.33 | 24.95 | 24.95 |
| 8 | 18.86 | 19.02 | 19.02 | 23.99 | 24.29 | 24.29 |
| 9 | 18.71 | 18.80 | 18.80 | 23.78 | 23.95 | 23.95 |
| 10 | 18.61 | 18.67 | 18.67 | 23.64 | 23.75 | 23.75 |
| 15 | 18.42 | 18.43 | 18.43 | 23.37 | 23.39 | 23.39 |
| 20 | 18.36 | 18.37 | 18.37 | 23.39 | 23.30 | 23.30 |
| ∞ | 18.31 | 18.31 | 18.31 | 23.21 | 23.21 | 23.21 |

*These numbers are outside the interval of convergence of (3.1) with $m = \infty$.

A. K. Gupta
Department of Mathematics and Statistics
Bowling Green State University
Bowling Green, Ohio 43403-0221

J. Tang
Bell Communications Research
6 Corporate Place
Piscataway, New Jersey 08854

## REFERENCES

Anderson, T. W. (1984). *An Introduction to Multivariate Statistical Analysis*, 2nd edition. Wiley, New York.

Box, G. E. P. (1949). 'A General Distribution Theory for a Class of Likelihood Criteria.' *Biometrika*, **36**, 317-346.

Gupta, A. K. (1977). 'On the Distribution of Sphericity Test Criterion in the Multivariate Gaussian Distribution.' *Aust. J. Statist.*, **19**, 202-205.

Gupta, A. K., and Tang, J. (1984). 'Distribution of Likelihood Ratio Statistic for Testing Equality of Covariance of Multivariate Gaussian Models.' *Biometrika*, **71**, 555-559.

Mathai, A. M., and Katiyar, R. S. (1979). 'Exact Percentage Points for Testing Independence.' *Biometrika*, **66**, 353-356.

Mudholkar, G. S., Trivedi, M. C., and Lin, C. T. (1982). 'An Approximation to the Distribution of Likelihood Ratio Statistic for Testing Complete Independence.' *Technometrics*, **24**, 139-143.

Tang, J., and Gupta, A. K. (1984). 'On the Distribution of the Product of Independent Beta Random Variables.' *Statist. & Probl. Letters*, **2**, 165-168.

Tang, J., and Gupta, A. K. (1987). 'On the Type-B Integral Equation and the Distribution of Wilks' Statistic for Testing Independence of Several Groups of Variables.' *Statistics*, **18**, (to appear).

Walster, G. W., and Tretter, M. J. (1980). 'Exact Noncentral Distribution of Wilks' $\Lambda$ and Wilks-Lawley $U$ Criteria as Mixtures of Incomplete Beta Functions: For Three Tests.' *Ann. Statist.*, **8**, 1388-1390.

Wilks, S. S. (1932). 'Certain Generalizations in Analysis of Variances.' *Biometrika*, **24**, 471-494.

D. R. Jensen

# TOPICS IN THE ANALYSIS OF REPEATED MEASUREMENTS

## ABSTRACT

This study is concerned with the analysis of repeated scalar measurements having $r \geq 1$ repetitions within cells of a two-way array. Alternative models are considered for dependencies among observations within subjects, and analytical methods are identified as appropriate for each. Procedures for multiple comparisons, for analyzing factorial experiments, and for other nonstandard tests are featured. Emphasis is given to the validity and efficiency of the several procedures considered. Nonparametric and robust aspects of relevant normal-theory tests are discussed with reference to the analysis of repeated measurements.

**Key words and phrases.** Repeated measurements, multiple comparisons, factorial experiments, validity, Hotelling's $T^2$, efficiency.

## 1. INTRODUCTION

Repeated measurements designs are used widely in medical trials and other experiments utilizing human and animal subjects, and elsewhere. In these studies $n$ experimental subjects are observed under different treatments on eack of $k$ successive occasions, the $i$th subject yielding the observation $y_{ij}$ on occasion $j$. The vector $\underset{\sim}{y}_i = [y_{i1}, y_{i2}, \dots, y_{ik}]'$ contains all records for subject $i$, and $\underset{\sim}{\mu} = [\mu_1, \mu_2, \dots, \mu_k]'$ contains the corresponding treatment means, the primes denoting transposition. The array $\underset{\sim}{Y} = [y_{ij}]$ constitutes a complete two-way classification with one observation per cell, and the problem is to test the hypothesis $H_0: \mu_1 = \mu_2 = \cdots = \mu_k$ against general alternatives. The problem, however, typically is complicated by dependencies among observations within subjects. These dependencies often have been overlooked (*cf.* Jennings and Wood (1976)), and it is now known that they assume a critical role in the analysis. Several approaches to the analysis of such experiments have been taken. These include (*i*) using the usual $F$ tests under types of dependencies shown to validate their use; (*ii*) using various approximate procedures for testing $H_0$ and for making multiple comparisons among the means; (*iii*) using multivariate methods requiring that $n > k$ but otherwise placing no constraints on



*H. Bozdogan and A. K. Gupta (eds.), Multivariate Statistical Modeling and Data Analysis, 147–161.*

the within-subject dispersion matrix $\underset{\sim}{\Sigma}$; and (*iv*) using methods for analyzing growth curves, including randomization analysis. References for approaches (*iii*) and (*iv*) are Morrison (1976) and Foutz *et al.* (1985), for example; further references are supplied subsequently. In this paper we reconsider approaches (*i*) and (*iii*) as they bear on some nonstandard topics in the analysis of repeated measurements. Specifically, we consider (*a*) multiple comparisons and the analysis of repeated measurements having treatments in factorial combination, (*b*) the analysis of repeated measurements having $r > 1$ repetitions within cells, and (*c*) the robustness of these procedures to normality assumptions.

To fix ideas, let $\underset{\sim}{1}_k = [1, 1, \dots, 1]'$ be the $k$-dimensional unit vector; let $\underset{\sim}{I}_k$ be the $(k \times k)$ identity matrix; and designate by $F(m, \nu, \lambda)$ and $T^2(m, \nu, \lambda)$ the Snedecor-Fisher and the Hotelling-Mahalanobis distributions having the noncentrality parameter $\lambda$ and error degrees of freedom $\nu$. Suppose that $\{\underset{\sim}{y}_1, \underset{\sim}{y}_2, \dots, \underset{\sim}{y}_n\}$ are independent $k$-dimensional Gaussian vectors all having the means $[\mu_1, \mu_2, \dots, \mu_k]'$ and the $(k \times k)$ dispersion matrix $\underset{\sim}{\Sigma}$, and let $\bar{\underset{\sim}{y}} = [\bar{y}_1, \bar{y}_2, \dots, \bar{y}_k]'$ be the sample means and $\underset{\sim}{S} = (n-1)^{-1}\sum_{i=1}^{n}(\underset{\sim}{y}_i - \bar{\underset{\sim}{y}})(\underset{\sim}{y}_i - \bar{\underset{\sim}{y}})'$ the sample dispersion matrix. The univariate test for $H_0$ uses the statistic

$$F = n\bar{\underset{\sim}{y}}'\underset{\sim}{C}(\underset{\sim}{C}'\underset{\sim}{C})^{-1}\underset{\sim}{C}'\bar{\underset{\sim}{y}}/(k-1)s^2 \tag{1.1}$$

where $s^2$ is the *Subject* × *Treatment* mean square having $\nu = (n-1)(k-1)$ degrees of freedom. Here the rows of $\underset{\sim}{C}'$ $(q \times k)$ are any set of $q = k - 1$ linearly independent contrasts on the Euclidean space $R^k$. This test is valid if and only if $\underset{\sim}{\Sigma}$ has the structure $\underset{\sim}{\Sigma} = \underset{\sim}{\Sigma}(\gamma)$ (*cf.* Huynh and Feldt (1970) and Rouanet and Lépine (1970)), *i.e.*,

$$\underset{\sim}{\Sigma}(\gamma) = [\gamma_i + \gamma_j + \gamma\delta_{ij}], \quad \gamma > 0 \tag{1.2}$$

where $\delta_{ij} = 1$ when $i = j$ and is zero otherwise. This is called the *sphericity* condition, under which sets of orthonormal linear contrasts are spherical Gaussian variables. A special case of (1.2) is the model for compound symmetry in which $\underset{\sim}{\Sigma}(\rho) = \sigma^2\{(1-\rho)\underset{\sim}{I}_k + \rho\underset{\sim}{1}_k\underset{\sim}{1}_k'\}$, long known to be sufficient for validity. The multivariate test for $H_0$, valid for any $\underset{\sim}{\Sigma}$ when $n > k$, uses the statistic

$$T^2 = n\bar{\underset{\sim}{y}}'\underset{\sim}{C}(\underset{\sim}{C}'\underset{\sim}{S}\underset{\sim}{C})^{-1}\underset{\sim}{C}'\bar{\underset{\sim}{y}}. \tag{1.3}$$

An approximate test for use when $\underset{\sim}{\Sigma} \neq \underset{\sim}{\Sigma}(\gamma)$, due to Geisser and Greenhouse (1958), uses the statistic (1.1) together with empirical adjustments to the degrees of freedom attributed to it.

The structure of $\underset{\sim}{\Sigma}$ is critical to the analysis. Several studies have shown that even small departures from sphericity can seriously affect the size and power of the

$F$ test; see Maxwell (1980) and Boik (1981) for supporting evidence and further references. Moreover, repeated measurements in education, psychology, and medicine often do not exhibit sphericity; see Keselman and Rogan (1981) and Keselman and Keselman (1984) for further details. Various empirical studies indicate that the actual level of the empirically adjusted test of Geisser and Greenhouse (1958) is slightly less than $\alpha$, and of the unadjusted test is greater than $\alpha$, in using the statistic (1.1) at the nominal level $\alpha$ when $\underset{\sim}{\Sigma} \neq \underset{\sim}{\Sigma}(\gamma)$; see Boik (1981), for example. Further empirical studies of Rogan *et al.* (1979) suggest that the multivariate $T^2$ test and the empirically adjusted $F$ test of Geisser and Greenhouse (1958) are comparable with respect to level and power.

In view of the alternative designs and analyses available for two-way experiments, it is of interest to study their comparative efficiencies. Design efficiencies have been studied for $F$ tests under two types of designs, the standard designs using $kn$ different subjects, for which $\underset{\sim}{\Sigma} = \underset{\sim}{\Sigma}(0) = \sigma^2 \underset{\sim}{I}_k$, and alternative designs using repeated measurements having the structure $\underset{\sim}{\Sigma}(\gamma)$ or $\underset{\sim}{\Sigma}(\rho)$. For testing the hypothesis $H_0: \mu_1 = \mu_2 = \cdots = \mu_k$ against general alternatives, exact small-sample relative efficiencies, $E\{\cdot, \cdot\}$, of the $F$ test under two designs may be found as ratios of the noncentrality parameters. These are listed in Table I from Jensen (1982), where the argument for $F(\cdot)$ identifies the structure of $\underset{\sim}{\Sigma}(\cdot)$. These results, which hold uniformly at all alternatives, will be seen to apply more generally. The $T^2$ test is based on fewer degrees of freedom for error than the $F$ test, and some authors have concluded erroneously that $T^2$ necessarily has less power. Indeed, this often is used to support the approximate procedures of Geisser and Greenhouse (1958). To clarify this point, asymptotic efficiencies of Hotelling's $T^2$ relative to the $F$ test have been studied for the special case that $\{\sigma_{11} = \sigma_{22} = \cdots = \sigma_{kk} = \sigma^2\}$ and thus $\underset{\sim}{\Sigma} = \sigma^2 \underset{\sim}{R}$, where $\underset{\sim}{R}$ is a correlation matrix; see Jensen (1982). Let $\{\omega_1 \geq \omega_2 \geq \cdots \geq \omega_q\}$ be the ordered characteristic values of $\underset{\sim}{\Omega} = (\underset{\sim}{I}_k - \underset{\sim}{R})\underset{\sim}{C}(\underset{\sim}{C}'\underset{\sim}{C})^{-1}\underset{\sim}{C}'$, where $q = k - 1$ and $\underset{\sim}{C}'$ is defined as before. Then the bounds

$$(1 - \omega_q)^{-1} \leq E\{T^2, F(0)\} \leq (1 - \omega_1)^{-1} \tag{1.4}$$

hold uniformly at all alternatives to $H_0$, and the bounds

$$1/2 < E\{T^2, F(0)\} < \infty \tag{1.5}$$

hold uniformly for all values of $\sigma^2 \underset{\sim}{R}$ as well. Although these results are asymptotic, the conclusions are essentially unchanged when adjusted for the fewer error degrees of freedom for $T^2$; see Jensen (1982).

## 2. LINEAR CONTRASTS

### 2.1 Basic Properties

We state essential properties of contrasts that are central to validating the use of conventional normal-theory procedures under specified dependencies. To these ends arrange the elements of $\underset{\sim}{Y} = [y_{ij}]$ by rows into the data array $\underset{\sim}{y} = [y_1', y_2', \ldots, y_n']'$, of order $(nk \times 1)$, and observe that its dispersion matrix is the Kronecker product $\underset{\sim}{I}_n \times \underset{\sim}{\Sigma}$. The following theorem provides a rather complete account of the effects of certain dependencies on the analysis of variance and related tests in repeated measurements experiments.

**Theorem 1.** Let $\{y_1, y_2, \ldots, y_n\}$ be independent Gaussian $k$-vectors all having the mean $\underset{\sim}{\mu}$ and the dispersion matrix $\underset{\sim}{\Sigma}$. Further let $\{\underset{\sim}{y}'\underset{\sim}{A}_0\underset{\sim}{y}, \underset{\sim}{y}'\underset{\sim}{A}_1\underset{\sim}{y}, \ldots, \underset{\sim}{y}'\underset{\sim}{A}_m\underset{\sim}{y}\}$ be quadratic forms in $\underset{\sim}{y}$ such that $s^2 = \underset{\sim}{y}'\underset{\sim}{A}_0\underset{\sim}{y}/(n-1)(k-1)$ as in expression (1.1).

(i) Every set of linear contrasts among the elements of $\bar{\underset{\sim}{y}} = [\bar{y}_1, \bar{y}_2, \ldots, \bar{y}_k]'$ under the dispersion structure $\underset{\sim}{\Sigma}(\underset{\sim}{\gamma})$ has properties identical to those under $\underset{\sim}{\Sigma}(0) = \sigma^2\underset{\sim}{I}_k$.

(ii) If $\{\underset{\sim}{y}'\underset{\sim}{A}_1\underset{\sim}{y}, \ldots, \underset{\sim}{y}'\underset{\sim}{A}_m\underset{\sim}{y}\}$ are sums of squares for sets of contrasts among the elements of $\bar{\underset{\sim}{y}} = [\bar{y}_1, \bar{y}_2, \ldots, \bar{y}_k]'$, then the Fisher-Cochran theorem is satisfied by the quadratic forms $\{\underset{\sim}{y}'\underset{\sim}{A}_0\underset{\sim}{y}, \underset{\sim}{y}'\underset{\sim}{A}_1\underset{\sim}{y}, \ldots, \underset{\sim}{y}'\underset{\sim}{A}_m\underset{\sim}{y}\}$ under the dispersion structure $\underset{\sim}{\Sigma}(\underset{\sim}{\gamma})$ if and only if it is satisfied under $\underset{\sim}{\Sigma}(\underset{\sim}{0}) = \sigma^2\underset{\sim}{I}_k$.

**Proof.** Let the rows of $\underset{\sim}{C}'$ $(q \times k)$ be a fixed set of $q = k - 1$ linearly independent contrasts. Because every set of contrasts among the elements of $\bar{\underset{\sim}{y}} = [\bar{y}_1, \bar{y}_2, \ldots, \bar{y}_k]'$ can be represented in terms of $\underset{\sim}{C}'\bar{\underset{\sim}{y}}$, it suffices to consider the latter. From the expression $\underset{\sim}{\Sigma}(\underset{\sim}{\gamma}) = (\underset{\sim}{\gamma}\underset{\sim}{1}_k' + \underset{\sim}{1}_k\underset{\sim}{\gamma}' + \gamma\underset{\sim}{I}_k)$ and the fact that $\underset{\sim}{C}'\underset{\sim}{1}_k = \underset{\sim}{0}$, where $\underset{\sim}{\gamma} = [\gamma_1, \gamma_2, \ldots, \gamma_k]'$, we infer that the dispersion matrix of $\underset{\sim}{C}'\bar{\underset{\sim}{y}}$ is proportional to

$$\underset{\sim}{C}'\underset{\sim}{\Sigma}(\underset{\sim}{\gamma})\underset{\sim}{C} = \underset{\sim}{C}'(\underset{\sim}{\gamma}\underset{\sim}{1}_k' + \underset{\sim}{1}_k\underset{\sim}{\gamma}' + \gamma\underset{\sim}{I}_k)\underset{\sim}{C} = \gamma\underset{\sim}{C}'\underset{\sim}{I}_k\underset{\sim}{C} \tag{2.1}$$

precisely as in the case $\underset{\sim}{\Sigma}(0) = \sigma^2\underset{\sim}{I}_k$, thus proving conclusion (*i*). To establish conclusion (*ii*) we apply standard tests for the independence and chi-squared ($\chi^2$) character of quadratic forms in the elements of $\underset{\sim}{y} = [y_1', y_2', \ldots, y_n']'$ having the dispersion matrix $\underset{\sim}{I}_n \times \underset{\sim}{\Sigma}$. Independence requires that $\underset{\sim}{A}_i(\underset{\sim}{I}_n \times \underset{\sim}{\Sigma})\underset{\sim}{A}_j = \underset{\sim}{0}$ for $i \neq j$. Apart from scaling, the chi-squared character of such forms is assured whenever $\underset{\sim}{A}_i(\underset{\sim}{I}_n \times \underset{\sim}{\Sigma})\underset{\sim}{A}_i = \kappa\underset{\sim}{A}_i$ for some $\kappa > 0$. Now assuming the structure $\underset{\sim}{\Sigma} = \underset{\sim}{\Sigma}(\underset{\sim}{\gamma})$ and using properties of $\{\underset{\sim}{y}'\underset{\sim}{A}_1\underset{\sim}{y}, \ldots, \underset{\sim}{y}'\underset{\sim}{A}_m\underset{\sim}{y}\}$ as sums of squares for sets of contrasts among the elements of $\bar{\underset{\sim}{y}} = [\bar{y}_1, \bar{y}_2, \ldots, \bar{y}_k]'$, we see that conclusion (*ii*) now follows along the lines of the proof for conclusion (*i*).

## 2.2 Multiple Comparisons

Multiple comparisons are central to modern data analysis. Various procedures apply routinely in the normal-theory analysis of two-way data when $\underset{\sim}{\Sigma} = \sigma^2 \underset{\sim}{I}_k$; a standard reference is Miller (1966). In particular, pair-wise comparisons of means are done routinely using procedures due to Tukey, Duncan, and Newman-Keuls, for example. Pairwise comparisons with a control follow Dunnett's method. A critical question is whether or not these multiple comparisons carry over with the $F$ test to the model $\underset{\sim}{\Sigma} = \underset{\sim}{\Sigma}(\gamma)$. A routine application of conclusion (*i*) of Theorem 1 assures affirmatively that these procedures, and every other normal-theory procedure based on contrasts among the elements of $\underset{\sim}{\bar{y}} = [\bar{y}_1, \bar{y}_2, \ldots, \bar{y}_k]'$, continue to hold exactly under the dispersion structure $\underset{\sim}{\Sigma}(\gamma)$.

These facts have been noted earlier in special cases. Multiple comparisons using a pooled error term are known to be sensitive to circularity (*cf.* Keselman *et al.* (1981)). In particular, Tukey's test is sensitive to various assumptions (*cf.* Keselman and Rogan (1977)), and its type 1 error rate in repeated measurements is inflated when sphericity fails (*cf.* Maxwell (1980)). Various pair-wise procedures have been advocated when the validating structure $\underset{\sim}{\Sigma} = \underset{\sim}{\Sigma}(\gamma)$ fails. Maxwell (1980) concluded from empirical studies that, of five procedures considered, only those using a separate error term for each comparison controlled the type 1 error rate at $\alpha$. Of these, Maxwell recommended an approximate Bonferroni method. Regardless of the structure of $\underset{\sim}{\Sigma}$, exact but conservative procedures are available based on Hotelling's $T^2$ projections; see Morrison (1976), page 147, for example. From empirical studies Boik (1981) concluded that these should be employed for posterior testing in repeated measurements.

The efficiency comparisons of Table I carry over in a natural way to multiple comparisons *via* simultaneous confidence bounds. Specifically, if relative efficiencies are gauged instead in terms of the expected lengths of confidence intervals under alternative models, then the ratios given in Table I apply directly. To be precise, let $q_\alpha(k, \nu)$ be the upper $100\alpha$ percentage point of the studentized range distribution having parameters $(k, \nu)$, *i.e.*, the distribution of the range of $k$ *iid* $N(0,1)$ variables divided by the square root of an independent $(\chi^2_\nu/\nu)$ variate having $\nu$ degrees of freedom. See Miller (1966). Simultaneous confidence bounds for the set of all pair-wise differences among means are given by

$$P(\mu_i - \mu_j \in \bar{y}_i - \bar{y}_j \pm q_\alpha(k, \nu)(s^2/n)^{1/2} : i \neq j) = 1 - \alpha \tag{2.2}$$

where $[\bar{y}_1, \bar{y}_2, \ldots, \bar{y}_k]'$ are means of n observations each. Similarly for $\underset{\sim}{c}' = [c_1, c_2, \ldots, c_k]$ belonging to the space $L_c$ of linear contrasts on $R^k$, *i.e.*, $L_c = \{\underset{\sim}{c} \in R^k : \sum_{i=1}^{k} c_i = 0\}$, the simultaneous confidence bounds for contrasts are

$$P(\underset{\sim}{c}'\underset{\sim}{\mu} \in \underset{\sim}{c}'\bar{\underset{\sim}{y}} \pm q_\alpha(k, \nu)(s^2/n)^{1/2} \sum_{i=1}^{k} |c_i| : \underset{\sim}{c} \in L_c) = 1 - \alpha. \tag{2.3}$$

Under alternative models for dependence the ratios of expected lengths of confidence intervals are given by the square roots of entries appearing in Table I. For example, confidence intervals are shorter on average by a factor $(1 - \rho)^{1/2}$ under the model $\underset{\sim}{\Sigma}(\rho)$ than under $\underset{\sim}{\Sigma}(0)$ when $\rho > 0$.

In lieu of normal-theory analyses, multiple comparisons also may be carried out through randomization analysis; see Foutz *et al.* (1985). It should be noted that these carry no special requirements regarding the within-subject dispersion matrix $\underset{\sim}{\Sigma}$ or the underlying distribution, so that much of the present discussion has no bearing on validity of randomization procedures.

## 2.3 Factorial Experiments

In practice it is not uncommon for the $k$ treatments to be in balanced factorial combination. Little has been settled regarding the analysis of such experiments in the case of repeated measurements. It is relevant to inquire whether the usual analysis for factorial experiments carries forward under suitable dependencies, and what types of analyses otherwise might be appropriate. To fix ideas suppose that there are two factors, $A$ and $B$, at levels $a$ and $b$ such that $k = ab$. Consider null hypotheses $H_A$, $H_B$, and $H_{AB}$ regarding the main effects of $A$ and $B$ and the $A \times B$ interaction. With reference to these let $\underset{\sim}{C}_A'$, $\underset{\sim}{C}_B'$, and $\underset{\sim}{C}_{AB}'$ be linear contrasts among the elements of $\bar{\underset{\sim}{y}} = [\bar{y}_1, \bar{y}_2, \dots, \bar{y}_k]'$ giving these main effects and interaction; represent by the quadratic forms $\{\underset{\sim}{y}'\underset{\sim}{A}_0\underset{\sim}{y}, \underset{\sim}{y}'\underset{\sim}{A}_1\underset{\sim}{y}, \underset{\sim}{y}'\underset{\sim}{A}_2\underset{\sim}{y}, \underset{\sim}{y}'\underset{\sim}{A}_3\underset{\sim}{y}\}$ the sums of squares due to *Error*, $A$, $B$, and $A \times B$ ; and let $F_A$, $F_B$, and $F_{AB}$ be corresponding versions of the statistic (1.1), where *Error* is the *Subjects* × *Treatments* interaction. The normal-theory analysis of balanced factorial experiments appropriate for the model $\underset{\sim}{\Sigma} = \sigma^2\underset{\sim}{I}_k$ remains exact in repeated measurements experiments having the structure $\underset{\sim}{\Sigma} = \underset{\sim}{\Sigma}(\gamma)$. This follows from conclusion (*i*) of Theorem 1. Specifically, the Fisher-Cochran theorem applies under $\underset{\sim}{\Sigma} = \underset{\sim}{\Sigma}(\gamma)$ to the quadratic forms $\{\underset{\sim}{y}'\underset{\sim}{A}_0\underset{\sim}{y}, \underset{\sim}{y}'\underset{\sim}{A}_1\underset{\sim}{y}, \underset{\sim}{y}'\underset{\sim}{A}_2\underset{\sim}{y}, \underset{\sim}{y}'\underset{\sim}{A}_3\underset{\sim}{y}\}$ by conclusion (*ii*); the joint distribution of the statistics $\{F_A, F_B, F_{AB}\}$ is precisely as given by Ghosh (1955); and Kimball's (1951) inequality applies.

Exact tests that are valid for any $\underset{\sim}{\Sigma}$ can be constructed using suitable versions of Hotelling's $T^2$ statistic when the sample size is sufficient. Let $T_A^2$, $T_B^2$, and $T_{AB}^2$ be as in expression (1.3) on replacing $\underset{\sim}{C}$ successively by $\underset{\sim}{C}_A'$, $\underset{\sim}{C}_B'$, and $\underset{\sim}{C}_{AB}'$. Then the hypotheses $H_A$, $H_B$, and $H_{AB}$ can be tested on referring $T_A^2$, $T_B^2$, and $T_{AB}^2$ to appropriate critical values of the central distribution $T^2(m, \nu, 0)$ with $\nu = n - 1$ and with $m$ successively taking the values $a - 1$, $b - 1$, and $(a - 1)(b - 1)$.

Alternative procedures have been studied by using the approach of Geisser and Greenhouse (1958) in which the degrees of freedom of the usual $F$ statistics

**Table I.**
Relative Design Efficiency $E\{F(\cdot), F(\cdot)\}$ of the Procedure of Row $i$ to Column $j$ under the $F$ Test.

| | $F(0)$ | $F(\rho)$ | $F(\gamma)$ |
|---|---|---|---|
| $F(0)$ | 1 | $(1 - \rho)$ | $\gamma/\sigma^2$ |
| $F(\rho)$ | $(1 - \rho)^{-1}$ | 1 | $\gamma/(1 - \rho)\sigma^2$ |
| $F(\gamma)$ | $\sigma^2/\gamma$ | $(1 - \rho)\sigma^2/\gamma$ | 1 |

**Table II.**
Details of Hypothesis Tests for the Data in Table III.

| Hypothesis | Test Statistic | Level | Critical Value |
|---|---|---|---|
| Sphericity | $-5 \ln W = 2.6063$ | 0.20 | $\chi^2_{0.20}(2) = 3.219$ |
| $H_0: \mu_1 = \mu_2 = \mu_3$ | $F = 18.78$ | 0.01 | $F_{0.01}(2, 12) = 6.93$ |
| $H_0: \mu_1 = \mu_2 = \mu_3$ | $T^2 = 48.83$ | 0.01 | $T^2_{0.01}(2, 6) = 31.857$ |

**Table III.**
Blood $CO_2$ of Rabbits under Three Respiration Schemes.

| Subject | Treatment 1 | | | Treatment 2 | | | Treatment 3 | | |
|---|---|---|---|---|---|---|---|---|---|
| 1 | 27.4 | 31.6 | 30.1 | 27.1 | 32.2 | 31.8 | 30.0 | 35.4 | 33.7 |
| 2 | 24.3 | 31.2 | 27.2 | 29.2 | 27.7 | 31.6 | 37.2 | 35.8 | 30.7 |
| 3 | 31.8 | 33.5 | 31.9 | 31.1 | 35.1 | 35.9 | 35.1 | 35.2 | 36.8 |
| 4 | 31.7 | 31.9 | 28.8 | 27.3 | 31.2 | 30.0 | 30.5 | 34.6 | 31.7 |
| 5 | 32.9 | 32.4 | 31.5 | 35.4 | 33.3 | 30.7 | 32.8 | 36.8 | 36.3 |
| 6 | 32.4 | 33.5 | 30.5 | 33.1 | 31.6 | 30.3 | 32.0 | 38.9 | 32.8 |
| 7 | 29.9 | 33.5 | 29.2 | 34.4 | 37.3 | 34.0 | 36.0 | 36.7 | 39.8 |

are modified empirically for each test. See Huynh (1978), Keselman and Rogan (1980), Keselman *et al.* (1981), and Keselman and Keselman (1984). This *ad hoc* approach at best is approximate, and it somewhat misses the point that the joint distribution of the test statistics is at issue in multiple inference. It is not known how empirical adjustments for the marginal statistics might affect the approximating joint distribution. Moreover, the need for such approximate procedures at best is less than compelling given the availability of exact procedures, especially in view of the fact that the $T^2$ test cannot be substantially less efficient than the $F$ test even when both apply; refer to expression (1.5) and, for further discussion, to Jensen (1982).

As before one may compare efficiencies of alternative procedures for testing the hypotheses $H_A$, $H_B$, and $H_{AB}$ under different designs and using alternative test procedures. Under different structures for dependence the relative efficiencies of Table I apply here to tests for each of $H_A$, $H_B$, and $H_{AB}$. Moreover, in comparing Hotelling's $T^2$ with the corresponding $F$ test for any of $H_A$, $H_B$, and $H_{AB}$, bounds of the type (1.4) and (1.5) apply. In short, for any of these hypotheses there are dependencies $\Sigma = \sigma^2 \underset{\sim}{R}$ for which the $T^2$ test using repeated measurements has arbitrarily greater efficiency against some alternatives than the corresponding $F$ test in a two-way experiment using $nk$ independent subjects.

## 3. REPETITIONS WITHIN CELLS

In balanced two-way experiments it is commonplace to have $r > 1$ repetitions within cells of the two-way table, where the pooled within-cell error mean square is used to estimate the common variance without assuming additivity of *Blocks* and *Treatments*. This protocol is used on occasion with repeated measurements, where the effects of treatments may be followed through time. However, little is known at present regarding the proper analysis of such experiments. In this section we undertake to study this matter. It will be seen that the usual analysis is inappropriate unless somewhat rigid requirements are met on the within-cell and between-cell structure of the within-subject dispersion matrices.

### 3.1 The Model

Consider a balanced experiment in which there are $n$ subjects, $k$ treatments, and $r > 1$ repetitions within each cell of the two-way array. On dropping the subscript, arrange the $rk$ observations for a typical subject in the partitioned form $\underset{\sim}{y}' = [\underset{\sim}{y}_1', \underset{\sim}{y}_2', \dots, \underset{\sim}{y}_k']$, where $\underset{\sim}{y}_j'$, of order $(1 \times r)$, consists of observations recorded under treatment $j$, *i.e.*,

$$\underset{\sim}{y}' = [y_1 \cdots y_r | y_{r+1} \cdots y_{2r} | \cdots | y_{r(k-1)+1} \cdots y_{rk}]. \tag{3.1}$$

More conventional notation is $y_{ijl}$, the $l$th observation in cell $(i,j)$. The corresponding vector of expected values is

$$\underset{\sim}{\theta}' = [\theta_1 \ldots \theta_r | \theta_{r+1} \ldots \theta_{2r} | \ldots | \theta_{r(k-1)+1} \cdots \theta_{rk}] \tag{3.2}$$

which in turn may be written as $\theta' = [\underset{\sim}{\theta}_1', \underset{\sim}{\theta}_2', \ldots, \underset{\sim}{\theta}_k']$. For the case that means are homogeneous within cells, expression (3.2) becomes

$$\underset{\sim}{\theta}' = [\mu_1 \ldots \mu_1 | \mu_2 \ldots \mu_2 | \ldots | \mu_k \ldots \mu_k], \tag{3.3}$$

these being means of the k treatments in the usual formulation of the model.

The dispersion matrix of $\underset{\sim}{y}$, again in partitioned form, generally may be written as

$$\underset{\sim}{\Sigma} = \begin{bmatrix} \underset{\sim}{\Sigma}_{11} & \underset{\sim}{\Sigma}_{12} & \cdots & \underset{\sim}{\Sigma}_{1k} \\ \underset{\sim}{\Sigma}_{21} & \underset{\sim}{\Sigma}_{22} & \cdots & \underset{\sim}{\Sigma}_{2k} \\ \cdots & \cdots & \cdots & \cdots \\ \underset{\sim}{\Sigma}_{k1} & \underset{\sim}{\Sigma}_{k2} & \cdots & \underset{\sim}{\Sigma}_{kk} \end{bmatrix} \tag{3.4}$$

where $\underset{\sim}{\Sigma}_{ij}$ is of order $(r \times r)$.

### 3.2 Structure for Dependence

Consider a single subject; results eventually will be pooled over subjects in the usual manner. In this section we suppose that the $(rk \times rk)$ matrix $\underset{\sim}{\Sigma}$ is a Kronecker product $\underset{\sim}{\Sigma} = \underset{\sim}{\Gamma} \times \underset{\sim}{\Xi}$ with $\Gamma$ of order $(k \times k)$ and $\underset{\sim}{\Xi}$ of order $(r \times r)$. For the present we take the cell means to be homogeneous as in (3.3). In order to examine properties of the pooled error mean square, let $\{\underset{\sim}{z}_1, \underset{\sim}{z}_2, \ldots, \underset{\sim}{z}_k\}$ be $r$-dimensional vectors of within-cell deviations from the observed cell means, *i.e.*, $\underset{\sim}{z}_j = (\underset{\sim}{I}_r - r^{-1}\underset{\sim}{1}_r\underset{\sim}{1}_r')\underset{\sim}{y}_j$ for $1 \leq j \leq k$. Write $\underset{\sim}{z} = [\underset{\sim}{z}_1', \underset{\sim}{z}_2', \ldots, \underset{\sim}{z}_k']'$. It follows that the pooled within-cell sum of squares can be written as $\underset{\sim}{z}'\underset{\sim}{z} = \underset{\sim}{y}'\underset{\sim}{A}\underset{\sim}{y}$, where $\underset{\sim}{A}$ is the idempotent matrix $\underset{\sim}{A} = \underset{\sim}{I}_k \times (\underset{\sim}{I}_r - r^{-1}\underset{\sim}{1}_r\underset{\sim}{1}_r')$.

In view of the fact that $\underset{\sim}{z}$ consists of linear contrasts among observations within cells, we infer from the earlier work that the sum of squares $\underset{\sim}{y}'\underset{\sim}{A}\underset{\sim}{y}$, when properly scaled, can have a chi-squared distribution if and only if $\underset{\sim}{\Xi}$ has the structure (1.2), say $\underset{\sim}{\Xi} = \underset{\sim}{\Xi}(\lambda) = [\lambda_i + \lambda_j + \lambda\delta_{i,j}]$. Upon evaluating the products $\underset{\sim}{A}[\underset{\sim}{\Gamma} \times \underset{\sim}{\Xi}(\lambda)]$ and $\underset{\sim}{A}[\underset{\sim}{\Gamma} \times \underset{\sim}{\Xi}(\lambda)]\underset{\sim}{A}$, recalling that $\underset{\sim}{A} = \underset{\sim}{I}_k \times (\underset{\sim}{I}_r - r^{-1}\underset{\sim}{1}_r\underset{\sim}{1}_r')$, we find that

$$\underset{\sim}{A}\,[\underset{\sim}{\Gamma} \times \underset{\sim}{\Xi}(\underset{\sim}{\lambda})] = \underset{\sim}{\Gamma} \times \left[\left(\underset{\sim}{\lambda} - \frac{\tau}{r}\underset{\sim}{1}_r\right)\underset{\sim}{1}_r' + \lambda\left(\underset{\sim}{I}_r - \frac{1}{r}\underset{\sim}{1}_r\underset{\sim}{1}_r'\right)\right] \tag{3.5}$$

$$\underset{\sim}{A}\,[\underset{\sim}{\Gamma} \times \underset{\sim}{\Xi}(\underset{\sim}{\lambda})]\,\underset{\sim}{A} = \underset{\sim}{\Gamma} \times \lambda\left(\underset{\sim}{I}_r - \frac{1}{r}\underset{\sim}{1}_r\underset{\sim}{1}_r'\right) \tag{3.6}$$

where $\lambda = [\lambda_1, \lambda_2, \ldots, \lambda_r]'$ and $\tau = \lambda_1 + \lambda_2 + \cdots + \lambda_r$.

Using standard arguments the expected value $E(\underset{\sim}{z}'\underset{\sim}{z}) = E(\underset{\sim}{y}'\underset{\sim}{A}\underset{\sim}{y})$ is found on evaluating the trace of (3.5). Similarly, the chi-squared character of $\underset{\sim}{y}'\underset{\sim}{A}\underset{\sim}{y}$ is determined, apart from scaling, from (3.6) *via* the standard criterion that $\underset{\sim}{A}\underset{\sim}{\Sigma}\underset{\sim}{A} = \kappa\underset{\sim}{A}$ for some $\kappa > 0$. Recall that $\underset{\sim}{y}'\underset{\sim}{A}\underset{\sim}{y}$ is pooled over cells within a typical subject, and that these eventually are to be pooled over subjects if appropriate. These developments support without further proof the following conclusions regarding the unbiased estimation of the variance, and the distributions of such estimates, as summarized in the next theorem.

**Theorem 2.** Consider a repeated measurements experiment with $k$ treatments and $r$ repetitions within cells such that the $(rk \times rk)$ dispersion matrix for a typical subject is $\underset{\sim}{\Sigma} = \underset{\sim}{\Gamma} \times \underset{\sim}{\Xi}$, and let $\underset{\sim}{y}'\underset{\sim}{A}\underset{\sim}{y}$ be the pooled within-cell sum of squares of deviations from observed cell means.

(*i*) If $\underset{\sim}{\Gamma}$ is any correlation matrix, then $E(\underset{\sim}{y}'\underset{\sim}{A}\underset{\sim}{y}) = k(r-1)\lambda$.

(*ii*) Under Gaussian assumptions the distribution of $\underset{\sim}{y}'\underset{\sim}{A}\underset{\sim}{y}$, apart from scaling, is a $\chi^2$ distribution having $k(r-1)$ degrees of freedom if and only if $\underset{\sim}{\Gamma} = \kappa\underset{\sim}{I}_k$ for some $\kappa > 0$.

Somewhat more general conclusions are supported by the foregoing developments. Specifically, the within-cell dispersion matrix $\underset{\sim}{\Xi}(\underset{\sim}{\lambda})$, heretofore assumed constant for all subjects and treatments, could now vary as long as the scalar parameter $\lambda$ is held constant. The conclusions of Theorem 2 remain intact. Thus a curious type of heterogeneity is permitted for which there is no parallel in the standard case for which $\underset{\sim}{\Xi} = \sigma^2\underset{\sim}{I}_r$.

We have sought conditions validating the usual analysis of variance for balanced two-way experiments having $r > 1$ repetitions within cells. We conclude that observations within cells may be dependent according to the structure $\underset{\sim}{\Xi}(\underset{\sim}{\lambda})$, but that independence among subjects and treatments is needed for validity. Circumstances supporting this model arise when $nk$ different subjects are used, and then $r$ repeated measurements are recorded for each.

### 3.3 Analysis of Cell Totals

We return to the problem of comparing $k$ treatment means using $kr$ repeated measurements on $n$ subjects with $r$ repetitions per cell. In view of the preceding section we anticipate that cell totals might be used under less restrictive models for within-subject dependencies. To these ends we replace the typical vector observation $\underset{\sim}{y}_{ij}$ within cell $(i,j)$ by the sum $\underset{\sim}{1}_r'\underset{\sim}{y}_{ij}$. For a typical subject the resulting $k$-dimensional vector of expected values is $[\underset{\sim}{1}_r'\underset{\sim}{\theta}_1, \underset{\sim}{1}_r'\underset{\sim}{\theta}_2, \ldots, \underset{\sim}{1}_r'\underset{\sim}{\theta}_k]$, and the $(k \times k)$ dispersion matrix is

$$\underset{\sim}{L}'\underset{\sim}{\Sigma}\underset{\sim}{L} = \begin{bmatrix} \underset{\sim}{1}_r'\underset{\sim}{\Sigma}_{11}\underset{\sim}{1}_r & \underset{\sim}{1}_r'\underset{\sim}{\Sigma}_{12}\underset{\sim}{1}_r & \cdots & \underset{\sim}{1}_r'\underset{\sim}{\Sigma}_{1k}\underset{\sim}{1}_r \\ \underset{\sim}{1}_r'\underset{\sim}{\Sigma}_{21}\underset{\sim}{1}_r & \underset{\sim}{1}_r'\underset{\sim}{\Sigma}_{22}\underset{\sim}{1}_r & \cdots & \underset{\sim}{1}_r'\underset{\sim}{\Sigma}_{2k}\underset{\sim}{1}_r \\ \cdots & \cdots & \cdots & \cdots \\ \underset{\sim}{1}_r'\underset{\sim}{\Sigma}_{k1}\underset{\sim}{1}_r & \underset{\sim}{1}_r'\underset{\sim}{\Sigma}_{k2}\underset{\sim}{1}_r & \cdots & \underset{\sim}{1}_r'\underset{\sim}{\Sigma}_{kk}\underset{\sim}{1}_r \end{bmatrix} \tag{3.7}$$

where $\underset{\sim}{L} = \underset{\sim}{I}_k \times \underset{\sim}{1}_r$.

Under the homogeneous model (3.3) the hypothesis of interest again is $H_0$: $\mu_1 = \mu_2 = \cdots = \mu_k$. We now proceed as before. If $\underset{\sim}{L}'\underset{\sim}{\Sigma}\underset{\sim}{L}$ has the structure $\underset{\sim}{\Sigma}(\gamma)$, which may be examined using a standard normal-theory test, then the $F$ test using the statistic (1.1) based on cell totals is valid, with $(k - 1)$ and $(n - 1)(k - 1)$ degrees of freedom. Alternatively, Hotelling's $T^2$ statistic (1.3) may be used requiring no special structure for $\underset{\sim}{L}'\underset{\sim}{\Sigma}\underset{\sim}{L}$. It is clear that the relative design efficiencies of Table I apply, as do the bounds on $E\{T^2, F(0)\}$ given in expressions (1.4) and (1.5). These conclusions all hold for the analysis of cell totals, whatever may be the structure of the within-cell dispersion matrices. In particular, if the $(rk \times rk)$ matrix $\underset{\sim}{\Sigma}$ has the structure $\underset{\sim}{\Sigma} = \underset{\sim}{\Gamma} \times \underset{\sim}{\Xi}$ considered in Section 3.2, then conditions validating the $F$ test based on cell totals are that $\underset{\sim}{\Gamma}$ have the structure (1.2), while $\underset{\sim}{\Xi}$ may be any positive definite $(r \times r)$ matrix.

We finally observe that these procedures support conclusions somewhat more general than those claimed. In fact, if the means within cells are not homogeneous, then the foregoing procedures using cell totals actually support tests for the more general linear hypothesis

$$H_G\text{: } \underset{\sim}{1}_r'\underset{\sim}{\theta}_1 = \underset{\sim}{1}_r'\underset{\sim}{\theta}_2 = \cdots = \underset{\sim}{1}_r'\underset{\sim}{\theta}_k. \tag{3.8}$$

Multiple comparisons and the analysis of factorial experiments using cell totals proceed as in Section 2, and the efficiency comparisons of Table I and of expressions (1.4) and (1.5) continue to apply.

## 4. DISCUSSION

Our developments thus far presuppose that observations are Gaussian. Even here it is seen that different methods apply under different models for dependence among observations within subjects. Here we consider some problems of implementation; we briefly survey robust aspects of the normal-theory procedures treated earlier; and we conclude with a numerical example.

### 4.1 Tests for Structure

In the analysis of variance using the $F$ statistic (1.1), in supporting multiple comparisons procedures, and in the analysis of factorial experiments, the validity of these procedures has been seen to depend critically on the structure $\underset{\sim}{\Sigma} = \underset{\sim}{\Sigma}(\gamma)$ of repeated measurements. The corresponding structure of $\underset{\sim}{L}'\underset{\sim}{\Sigma}\underset{\sim}{L}$ assumes an equally critical role in the analysis of cell totals as in Section 3.3. A standard normal-theory test for the hypothesis $H\colon \underset{\sim}{\Sigma} = \underset{\sim}{\Sigma}(\gamma)$ is available; see Morrison (1976), page 251, for example. Often it is recommended that this be used as a preliminary test to determine which of the subsequent tests for means should be employed.

Keselman *et al.* (1980) concluded from empirical studies that preliminary tests for sphericity often are not useful in practice. Boik (1981) pointed out that the test for sphericity is not very powerful against the small departures from sphericity that nonetheless may cause serious disturbances in the level and power of the $F$ test. We suggest a middle course. Evidence regarding the sphericity hypothesis is of value in practice. As a type 2 error apparently is the more serious, the power of the preliminary test perhaps can be enhanced sufficiently through the classical tradeoff between level and power by permitting a substantial increase in its level.

### 4.2 Robustness to Normality

Section 4 of Jensen (1982) treats robustness to normality of the $F$ and $T^2$ tests in the analysis of repeated measurements having one observation per cell. The alternative distributions constitute the class of all ellipsoidal distributions studied by Kelker (1970) and others, and various subclasses of these distributions. It was shown that $\alpha$-level normal-theory $F$ and $T^2$ tests remain exact at level $\alpha$ for all ellipsoidal distributions, including ellipsoidal Cauchy distributions not having first moments. It was shown further that efficiency comparisons among $F(0)$, $F(\rho)$, $F(\gamma)$, and $T^2$ continue to hold for an interesting subclass of these distributions.

The earlier results on robustness carry over to the problems considered here. In particular, test procedures used in the analysis of factorial experiments, in multiple comparisons, and in the analysis of cell totals as in Section 3.3, if exact at level $\alpha$ under normality, remain exact at level $\alpha$ for every ellipsoidal distribution. Supporting arguments are supplied in Jensen (1982). In particular, the entries in Table I, when interpreted as ratios of the squared expected lengths of confidence intervals of the types (2.2) and (2.3), continue to hold for all ellipsoidal distributions.

### 4.3 An Example

Blood carbon dioxide ($CO_2$) was measured on $n = 7$ rabbits under each of $k = 3$ treatments having $r = 3$ repetitions within cells of the two-way table. Treatments are identified as follows:

Treatment 1: Spontaneous Respiration
Treatment 2: Assisted Respiration
Treatment 3: Assisted Respiration with Halothane

Data from the experiment are recorded in Table III. In view of the findings reported in Section 3, we compare treatments using cell totals. For these totals the observed treatment means are $\bar{\underset{\sim}{y}} = [92.46\ 95.76\ 104.11]'$, and the sample dispersion matrix based on $\nu = 6$ degrees of freedom is

$$S = \begin{bmatrix} 27.2329 & 21.8662 & 7.1290 \\ 21.8662 & 46.6595 & 31.4290 \\ 7.1290 & 31.4290 & 26.9014 \end{bmatrix}$$

We first check whether the sphericity assumption is tenable in order to validate the usual $F$ test. Recall that earlier published evidence suggests that this test is not very powerful against small departures from sphericity. Following our suggestion in Section 4.1, we enhance the power by using the classical trade-off and test at the 0.20 level. Details of this test are provided in Table II, where we have applied the test described on page 251 of Morrison (1976). In particular, in Morrison's notation the statistic $W$ takes the value $W = 0.5938$, and the test statistic here is $-5 \ln W = 2.6063$. On accepting the sphericity hypothesis at the 0.20 level, we proceed to the usual $F$ test using the statistic (1.1) based on cell totals as summarized in Table II. The evidence suggests conclusively that there are differences in the treatment means associated with the three types of respiration.

In view of the caution cited about using a preliminary test for sphericity in Section 4.1, we also apply the $T^2$ test for the purpose of corroboration. Details are provided in Table II. This test, requiring no special structure for $\underset{\sim}{\Sigma}$, strongly supports our earlier conclusions that the treatments do indeed have different effects on the blood $CO_2$ under conditions studied in this experiment.

## ACKNOWLEDGEMENTS

The author is indebted to Professors Hamparsum Bozdogan and A. K. Gupta for organizing this Symposium, and to Professor Yoshio Takane for supplying several references.

Department of Statistics
Virginia Polytechnic Institute and State University
Blacksburg, Virginia, 24061

## REFERENCES

Boik, R. J. (1981). 'A Priori Tests in Repeated Measures Designs: Effects of Nonsphericity.' *Psychometrika* **46**, 241-255.

Foutz, R. V., Jensen, D. R. and Anderson, G. W. (1985). 'Multiple Comparisons in the Randomization Analysis of Designed Experiments with Growth Curve Responses.' *Biometrics* **41**, 29-37.

Geisser, S. and Greenhouse, S. W. (1958). 'An Extension of Box's Results on the Use of the $F$ Distribution in Multivariate Analysis.' *Ann. Math Statist.* **29**, 885-891.

Ghosh, M. N. (1955). 'Simultaneous Tests of Linear Hypotheses.' *Biometrika* **42**, 441-449.

Huynh, H. (1978). 'Some Approximate Tests for Repeated Measurement Designs. *Psychometrika* **43**, 161-175.

Huynh, H. and Feldt, L. S. (1970). 'Conditions Under Which Mean Square Ratios in Repeated Measurements Designs Have Exact $F$-Distributions.' *J. Amer. Statist. Assoc.* **65**, 1582-1589.

Jennings, J. R. and Wood, C. C. (1976). 'The $\varepsilon$-Adjusted Procedure for Repeated Measures Analyses of Variance.' *Psychophysiology* **13**, 277-278.

Jensen, D. R. (1982). 'Efficiency and Robustness in the Use of Repeated Measurements.' *Biometrics* **38**, 813-825.

Kelker, D. (1970). 'Distribution Theory of Spherical Distributions and a Location-Scale Parameter Generalization.' *Sankhya* **32A**, 419-430.

Keselman, H. J. and Keselman, J. C. (1984). 'The Analysis of Repeated Measures Designs in Medical Research.' *Statistics in Medicine* **3**, 185-195.

Keselman, H. J. and Rogan, J. C. (1977). 'The Tukey Multiple Comparison Test: 1953-1976.' *Psychological Bull.* **5**, 1050-1056.

Keselman, H. J. and Rogan, J. C. (1980). 'Repeated Measures *F* Tests and Psychophysiological Research: Controlling the number of false positives.' *Psychobiology* **17**, 499-503.

Keselman, H. J., Rogan, J. C. and Games, P. A. (1981). 'Robust Tests of Repeated Measures Means in Educational and Psychological Research.' *Educa. and Psychol. Measurement* **41**, 163-173.

Keselman, H. J., Rogan, J. C., Mendoza, J. L. and Breen, L. J. (1980). 'Testing the Validity Conditions of Repeated Measures *F* Tests.' *Psychological Bull.* **87**, 479-481.

Kimball, A. W. (1951). 'On Dependent Tests of Significance in the Analysis of Variance.' *Ann. Math. Statist.* **22**, 600-602.

Maxwell, S. E. (1980). 'Pairwise Multiple Comparisons in Repeated Measures Designs.' *J. Educational Statist.* **5**, 269-287.

Miller, R. G. (1966). *Simultaneous Statistical Inference.* McGraw-Hill, New York.

Morrison, D. F. (1976). *Multivariate Statistical Methods, Second Edition.* McGraw-Hill, New York.

Rogan, J. C., Keselman, H. J. and Mendoza, J. L. (1979). 'Analysis of Repeated Measurements.' *British J. Math. and Statist. Psychology* **32**, 269-286.

Rouanet, H. and Lépine, D. (1970). 'Comparison Between Treatments in a Repeated-Measurement Design: ANOVA and Multivariate Methods.' *British J. Math. and Statist. Psychology* **23**, 147-163.

Stanley L. Sclove

# METRIC CONSIDERATIONS IN CLUSTERING: IMPLICATIONS FOR ALGORITHMS

ABSTRACT

Given measurements on p variables for each of n individuals, aspects of the problem of clustering the individuals are considered. Special attention is given to models based upon mixtures of distributions, esp. multivariate normal distributions. The relationship between the orientation(s) of the clusters and the nature of the within-cluster covariance matrices is reviewed, as is the inadequacy of transformation to principal components based on the overall (total) covariance matrix of the whole (mixed) sample. The nature of certain iterative algorithms is discussed; variations which result from allowing different covariance matrices within clusters are studied.

Key words and phrases:
Cluster analysis, Mahalanobis distance, mixture model, isodata, k-means

## 1. INTRODUCTION

### 1.1. Overview of the Paper

In this paper certain ways of plotting data and certain aspects of the problem of clustering individuals will be discussed.

The kind of data treated results from observation of the same p variables for each of n individuals. Table I shows a typical multivariate dataset, consisting of p = 5 variables (age, systolic and diastolic blood pressure,



*H. Bozdogan and A. K. Gupta (eds.), Multivariate Statistical Modeling and Data Analysis, 163–186.*

Table I
Typical
Multivariate Dataset:
p = 5 variables
for each of n = 20 cases

| | Age | Sys | Dias | Wt | Ht |
|---|---|---|---|---|---|
| A | 44 | 124 | 80 | 190 | 70 |
| B | 35 | 110 | 70 | 216 | 73 |
| C | 41 | 114 | 80 | 178 | 68 |
| D | 31 | 100 | 80 | 149 | 68 |
| E | 61 | 190 | 110 | 182 | 68 |
| F | 61 | 130 | 88 | 185 | 70 |
| G | 44 | 130 | 94 | 161 | 68 |
| H | 58 | 110 | 74 | 175 | 67 |
| I | 52 | 120 | 80 | 144 | 66 |
| J | 52 | 120 | 80 | 130 | 67 |
| K | 52 | 130 | 80 | 162 | 69 |
| L | 40 | 120 | 90 | 175 | 68 |
| M | 49 | 130 | 75 | 155 | 66 |
| N | 34 | 120 | 80 | 156 | 74 |
| O | 37 | 115 | 70 | 151 | 65 |
| P | 63 | 140 | 90 | 168 | 74 |
| Q | 28 | 138 | 80 | 185 | 70 |
| R | 40 | 115 | 82 | 225 | 69 |
| S | 51 | 148 | 110 | 247 | 69 |
| T | 33 | 120 | 70 | 146 | 66 |

Source:
First 20 cases
of Table 2-2a
in Dixon and Massey (1969)

weight and height) observed for n = 20 individuals (adult men). One can consider the data as n points in p-space, i.e., the axes represent the p variables and the points represent the n individuals. When p is 2, this gives the ordinary "scatterplot."

The paper has been written not just for specialists in statistics; rather, an attempt has been made to write at the level of scientifically oriented people who have a knowledge of statistics at a level typically gained from an initial, one-year course.

In Section 2 a method of plotting data in a way which facilitates viewing the observations relative to their means is shown and discussed. In Section 3 linear transformations are discussed. In Section 4 some distance concepts are discussed. In Section 5 the algorithm developed and discussed in Sclove (1977) is illustrated. Section 6 is a discussion of some possible improvements in the algorithm.

### 1.2. Notation

It is necessary to introduce notation to distinguish vectors and matrices from scalars, and to distinguish between random and fixed (nonrandom) variables.

Scalar random variables are denoted by x, y, z, etc.
Nonrandom scalars are denoted by a, b, c, etc.
Fixed vectors are denoted by $\underline{a}$, $\underline{b}$, $\underline{c}$, etc.
Random vectors are denoted by $\underline{x}$, $\underline{y}$, $\underline{z}$, etc.

Here is some further notation to be used to distinguish matrices from vectors and scalars:

Random matrices are denoted by $\underline{X}$, $\underline{Y}$, $\underline{Z}$, etc.
Fixed matrices are denoted by $\underline{A}$, $\underline{B}$, $\underline{C}$, etc.

This notation is summarized in Table II. Some of the statistical notation used in the paper is listed below.

p the number of variables

n the number of individuals (sample size)

$\underline{X}$ the p-by-n data matrix

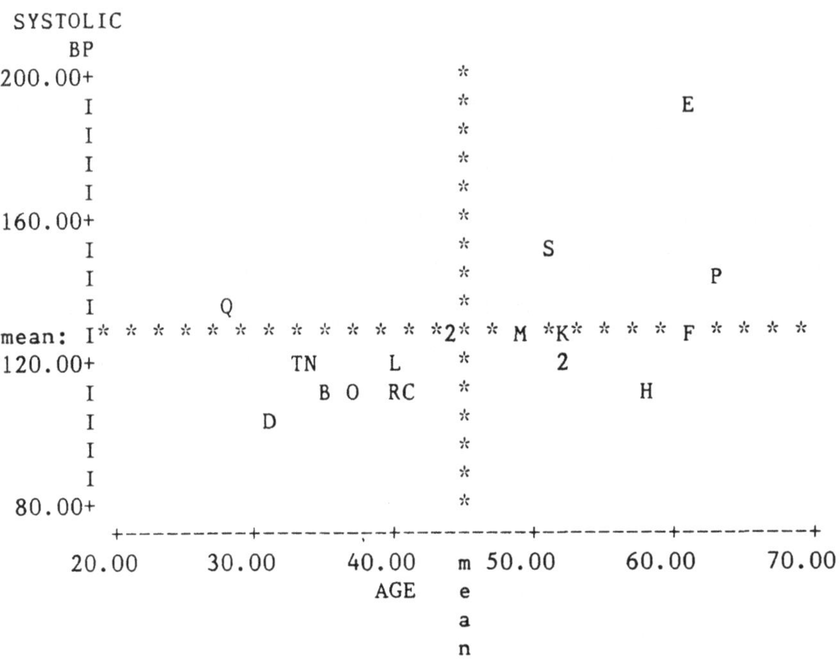


Figure 1
Plot of Systolic Blood Pressure (mean = 126.2)
vs. Age (mean = 45.3)
with Means Indicated by (*s)
for the 20 Patients in Table I

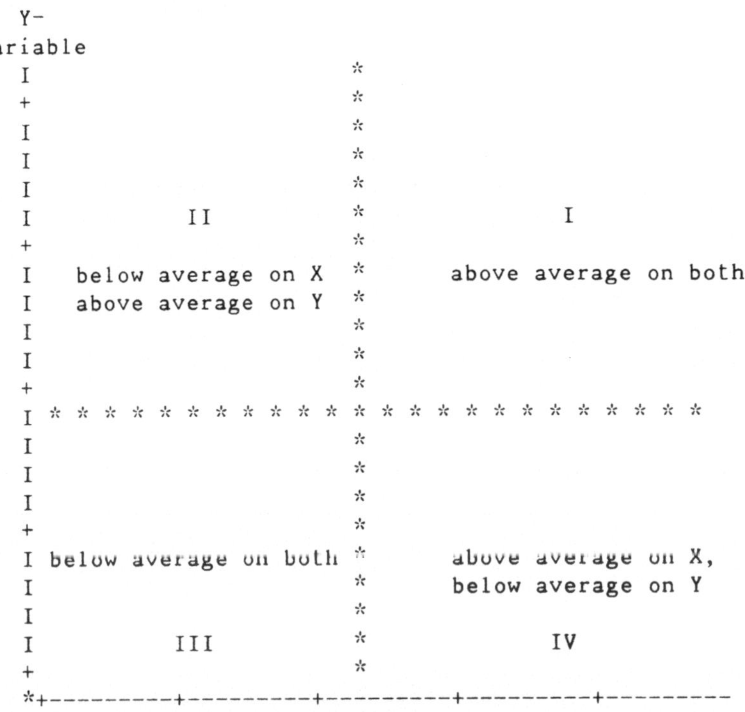


Figure 2
Axes Corresponding to the Means,
with Quadrants Indicated

Table II
Notation for Scalar, Vector and Matrix
Fixed and Random Variables

| | scalar (lower case) | vector (lower case, underlined) | matrix (upper case, underlined) |
|---|---|---|---|
| fixed (first part of alphabet) | a, b, c | $\underline{a}$, $\underline{b}$, $\underline{c}$ | $\underline{A}$, $\underline{B}$, $\underline{C}$ |
| random (latter part of alphabet) | x, y, z | $\underline{x}$, $\underline{y}$, $\underline{z}$ | $\underline{X}$, $\underline{Y}$, $\underline{Z}$ |

Table III
Values of Normalized Principal Components of Systolic and Diastolic Blood Pressure for the 20 Patients of Table I

| | NORMALIZED PC1 | NORMALIZED PC2 |
|---|---|---|
| A | 7.00 | 2.06 |
| B | 6.19 | 1.69 |
| C | 6.58 | 2.78 |
| D | 5.99 | 3.78 |
| E | 10.45 | 1.43 |
| F | 7.43 | 2.72 |
| G | 7.56 | 3.55 |
| H | 6.28 | 2.24 |
| I | 6.83 | 2.35 |
| J | 6.83 | 2.35 |
| K | 7.25 | 1.63 |
| L | 7.05 | 3.72 |
| M | 7.14 | .94 |
| N | 6.83 | 2.35 |
| O | 6.40 | 1.33 |
| P | 7.90 | 2.28 |
| Q | 7.59 | 1.05 |
| R | 6.66 | 2.98 |
| S | 8.68 | 4.45 |
| T | 6.61 | .97 |

| | |
|---|---|
| v | an index (subscript) denoting variables: $v = 1,2,\ldots,p$ |
| i | an index (subscript) denoting individuals: $i = 1,2,\ldots,n$ |
| $\underline{x}_i$ | the i-th column of the data matrix, $i = 1,2,\ldots,n$; i.e., the vector of scores of the i-th individual on all p variables |
| $\|\underline{M}\|$ | determinant of a given matrix $\underline{M}$ |
| $\underline{S}$ | p-by-p sample covariance matrix |
| $\underline{\Sigma}$ | p-by-p covariance matrix of the distribution (i.e., the "population" covariance matrix) |

## 2. PLOTTING WITH AXES GIVEN BY MEANS

Figure 1 is a plot of Systolic Blood Pressure vs. Age for the 20 patients of Table I. Note that most of the observations are in Quadrants I and III.

A general pattern for use in the statistical interpretation of a scatterplot is indicated in Figure 2.

The covariance can be interpreted in terms of this diagram. The contribution of Individual i to the covariance is

$$(x_i - \bar{x})(y_i - \bar{y}),$$

where the bars over x and y denote sample means. This is positive for individuals in Quadrants I and III and negative for those in Quadrants II and IV. The sample covariance is essentially the mean of these individual contributions. In summing over individuals either the positives or the negatives can predominate. If the plot is in terms of "z-scores,where the z-score of an individual on a variable is the number of standard deviations above or below the mean on that variable, then the covariance is the correlation coefficient. Thus, if most of the observations are in Quadrants I and III, it is indicative of positive correlation.

## 3. LINEAR TRANSFORMATIONS

Since several variables are observed for each individual, there is the possibility of replacing the original variables by functions of them.

Examples. (i) Having observed $p$ = price per share of stock and $e$ = last year's earnings per share for a number of firms, rather than analyzing $(p,e)$ one may study $(p,r)$, where $r = p/e$ is the familiar "price-earnings ratio." Of course, given $(p,r)$, one can recover the variable $e$, since $e = r/p$. (ii) The transformation from $(x,y,z)$ to $(s,x/s,y/s)$, where $s = x + y + z$, can be of interest for positive variables $x$, $y$, $z$. If $x$, $y$, $z$ are lengths, then the variable $s$ measures overall size and $x/s$, $y/s$ can be construed as shape variables. Note that the third ratio $z/s$ could be determined from the new variables $s$, $x/s$, $y/s$ from the relation $x/s + y/s + z/s = 1$. (iii) (continuation) The transformation from

$$(\log x, \log y, \log z) \text{ to } (t, \log x - t, \log y - t),$$

where $t = \log x + \log y + \log z$, can be of interest in this sort of situation. If $x$, $y$, $z$ are length, width and height, then the variable $t$ is log(volume). Note that this transformation is linear in the logarithms.

### 3.1. Linear Combinations

Often linear combinations (or linear combinations of transforms such as logarithms) are considered. If $p$ original variables are replaced by $p$ appropriately chosen linear combinations, all the information in the original variables is retained, in the sense that the original values could be recovered.

For example, if $x$ is systolic and $y$ is diastolic blood pressure and if instead of $x,y$, one records the sum

$$s = x + y$$

and the difference

$$d = x - y,$$

then, given for example that $s = 180$ and $d = 40$, one can recover the fact that $x = 120$ and $y = 70$.

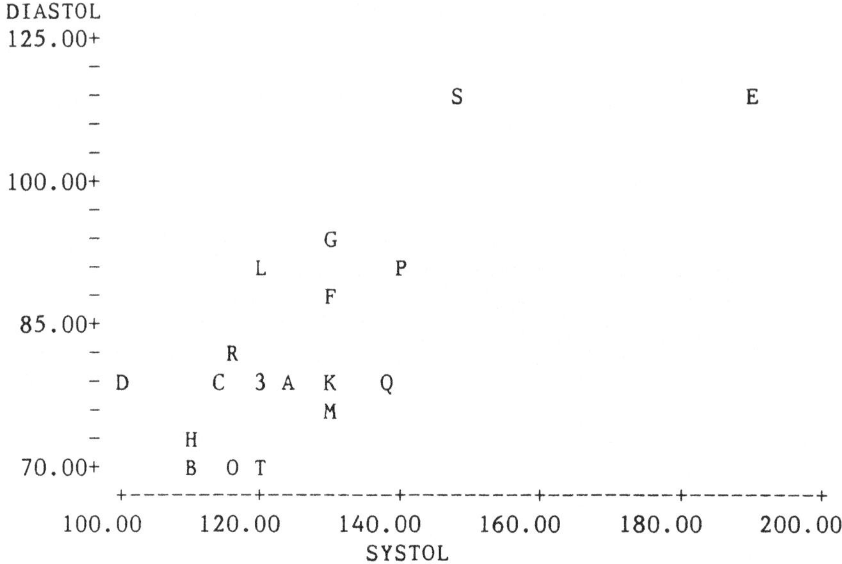


Figure 3
Diastolic vs. Systolic Blood Pressure
for the 20 Patients in Table I

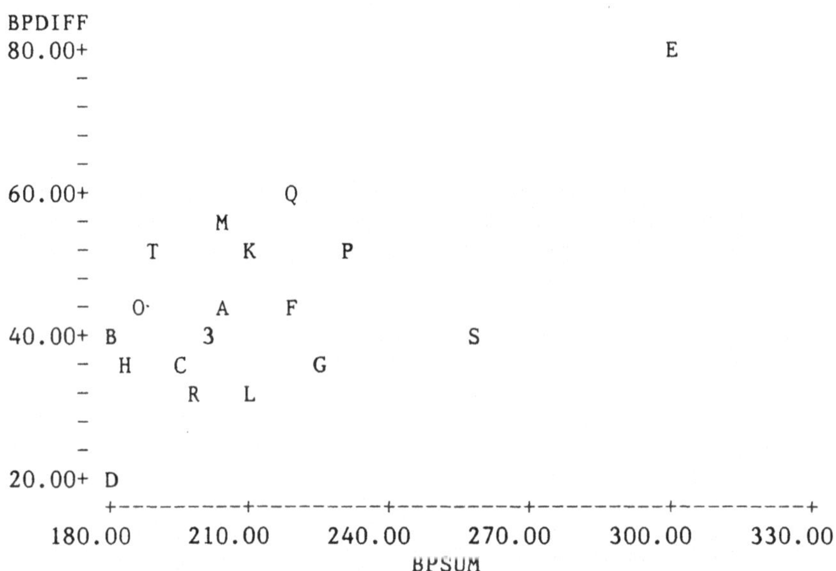


Figure 4
Difference vs. Sum of Systolic and Diastolic Blood Pressure

Figure 3 is a plot of diasolic vs. systolic blood pressure for the 20 patients of Table I. Figure 4 is a plot of the difference vs. the sum for these patients. Geometrically, Figure 4 is a rotation of Figure 3 by 45 degrees, with an expansion in both axes by a factor of the square root of 2.

If the variances of x and y are equal, as might be the case with scaled scores such as IQs, then their sum and difference are uncorrelated. Let us consider uncorrelated linear combinations more generally.

### 3.2. Uncorrelated Linear Combinations

To separate effects of different variables it is useful to transform to uncorrelated linear combinations of the original variables. There are many ways to do this. One way is to transform from, say,

$$(x,\ y,\ z) \quad \text{to} \quad (x,\ y - (y|x),\ z - (z|x,y)),$$

where $y|x$ denotes the regression of y on x and $z|x,y$ is the regression of z on x and y. More generally, given $\underline{x}$ with covariance matrix $\underline{\Sigma}$, let $\underline{z} = \underline{T}\ x$, where $\underline{T}$ is any matrix such that $\underline{T}\ \underline{\Sigma}\ \underline{T}'$, which is the covariance matrix of $\underline{z}$, is equal to a diagonal matrix, so that the p variables in $\underline{z}$ are uncorrelated. Since there are many ways to transform to uncorrelated linear combinations, we can insist that the transformation have further desirable properties. Perhaps the most interesting variables or combinations of variables are those with largest variance because they vary most across the individuals in the population. Hence, let us insist that the first several linear combinations retain as much information as possible in the sense of having the highest possible variance. This is perhaps especially appropriate when considering a dataset in which the variables have the same units of measurement, such as the blood pressure measurements, or IQ scores (which are scaled), or the Fisher iris data (Anderson 1935, 1936; Fisher, 1936), consisting of four measurements of length on each flower, namely petal and sepal length and width.
(which are scaled), or the Fisher iris data (Anderson 1935, 1936; Fisher, 1936), consisting of four measurements of

length on each flower, namely petal and sepal length and width.

### 3.3. Principal Components

For a mathematical discussion of principal components, see, e.g., Anderson (1984), or Johnson and Wichern (1982). It suffices here to discuss some aspects of principal components in a general way.

The "principal components" are uncorrelated linear combinations of maximal variance. That is, the first principal component (PC) is the linear combination having the largest variance (subject to a normalizing condition that the vector of coefficients have length one). Given a random vector $\underline{x}$ with covariance matrix $\underline{\Sigma}$, one measure of the size of the covariance matrix is its trace, the sum of its diagonal elements, which is simply the sum of the variances. In a sense, it is a "total variance." It is impossible to assess the information provided by a single variable because in general the variables are correlated, that is, the covariance matrix is not diagonal. If the variables were uncorrelated, their variances could be viewed as their contributions to total variance. We "load" as much variance as possible into the first PC. Then we load as much of the remaining variance as possible into the second PC. Etc.

In this way, we search for a few linear combinations which might be used to summarize the data. The first principal component is the best linear combination for representing the data in a single dimension. Recall that the "inner product" of two vectors, say (a,b,c) and (x,y,z), is the sum of products of corresponding elements, ax + by + cz. The first principal component, say PC1, turns out to be given by

PC1 = inner product of the observation vector $\underline{x}$ and $\underline{a}_1$,

the latter vector being the eigenvector corresponding to the largest eigenvalue of the covariance matrix. That is,

score of i-th individual on PC1 =
inner product of his observation vector with $\underline{a}_1$.

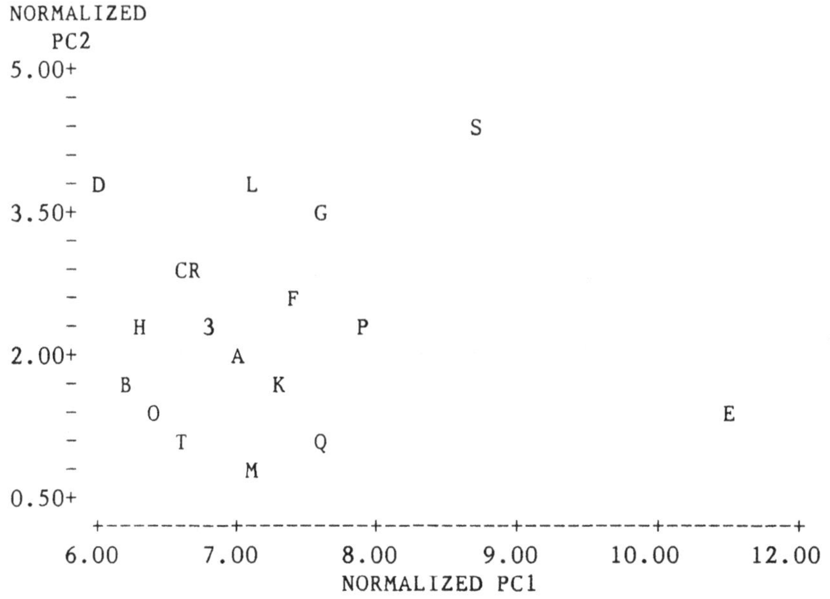


Figure 5
Plot of Principal Components
of Systolic and Diastolic Blood Pressures

An optimal way to plot the sample in a single dimension is to plot the scores on PC1. The percentage of total information retained in this one-dimensional representation is the ratio of the largest eigenvalue to the sum of all the eigenvalues, times 100%. Note that the sum of all the eigenvalues is simply the trace of $\underline{S}$, i.e., the "total variance." If this is large, say larger than 70% or 80%, a one-dimensional representation is adequate for most purposes. The second best linear combination say PC2, is the inner product of the observation vector with the eigenvector corresponding to the second largest eigenvalue. The proportion of information (total sample information) retained in a (PC1,PC2)-plot is

(sum of 2 largest eigenvalues/sum of all p eigenvalues).

Figure 5 is a plot of the 20 patients of Table I with respect to normalized PCs. These are the PCs, each divided by its standard deviation (the square root of the corresponding eigenvalue), so that the resulting scaled PCs have unit variance.

Note that all of the p original variables remain in the PCs, for each PC is a linear combination of all of the variables. However, for certain datasets some of the variables may have coefficients which are nearly zero and could be discarded without changing the values of the important PCs much.

The situation, then, is this, in terms of the sample covariance matrix $\underline{S}$. Let

$$\underline{a}_v \ , \quad v = 1,2,\ldots,p,$$

be p orthogonal eigenvectors of the sample covariance matrix $\underline{S}$. Then

score of i-th individual
on the v-th PC =
inner product of his observation vector with $\underline{a}_v$.

One can use axes corresponding to the first m PCs to plot the data in m-space ($m \leq p$). E.g., for an optimal plot of the the p-dimensional dataset in two-dimensional space (the plane), Individual i is represented by the point whose coordinates are his scores on PC1 and PC2. This is what has been done in Figure 5, with the modification that the PCs have been normalized to have variance one. The reason for this is a subject of the next section.

## 4. DISTANCE

It seems clear that in computing distances between individuals adjustments have to be made for correlations and for differing variances.

Euclidean distance (ordinary ruler distance) is a valid measure of distance when the p variables are uncorrelated and have equal variances. Thus Euclidean distance is a valid measure in Figure 5, a plot using normalized PC1 and PC2, uncorrelated variables of equal variance. The fact that Euclidean distance is a valid measure of distance when the p variables are uncorrelated tells us what is an appropriate measure of distance in general. For, let $\underline{z}$ denote a vector of uncorrelated variables and let $\underline{z1}$ and $\underline{z2}$ be observations of $\underline{z}$ for two individuals. Then the square of the Euclidean distance between these two individuals can be written as

$$(\underline{z1} - \underline{z2})'(\underline{z1} - \underline{z2}),$$

where the prime (') denotes vector transpose. Given $\underline{x}$ with covariance matrix $\underline{\Sigma}$, let $\underline{z} = \underline{T}\ \underline{x}$, where here $\underline{T}$ is any (nonsingular) matrix such that

$$\underline{T}\ \underline{\Sigma}\ \underline{T}' = \underline{I},$$

the identity matrix. Then the p variables in $\underline{z}$ are uncorrelated. The square of the distance between the two individuals is

$$\begin{aligned}
&(\underline{z1} - \underline{z2})'(\underline{z1} - \underline{z2}) \\
&= (\underline{T}\ \underline{x1} - \underline{T}\ \underline{x2})'(\underline{T}\ \underline{x1} - \underline{T}\ \underline{x2}) \\
&= (\underline{x1} - \underline{x2})'(\underline{T}'\underline{T})(\underline{x1} - \underline{x2}),
\end{aligned}$$

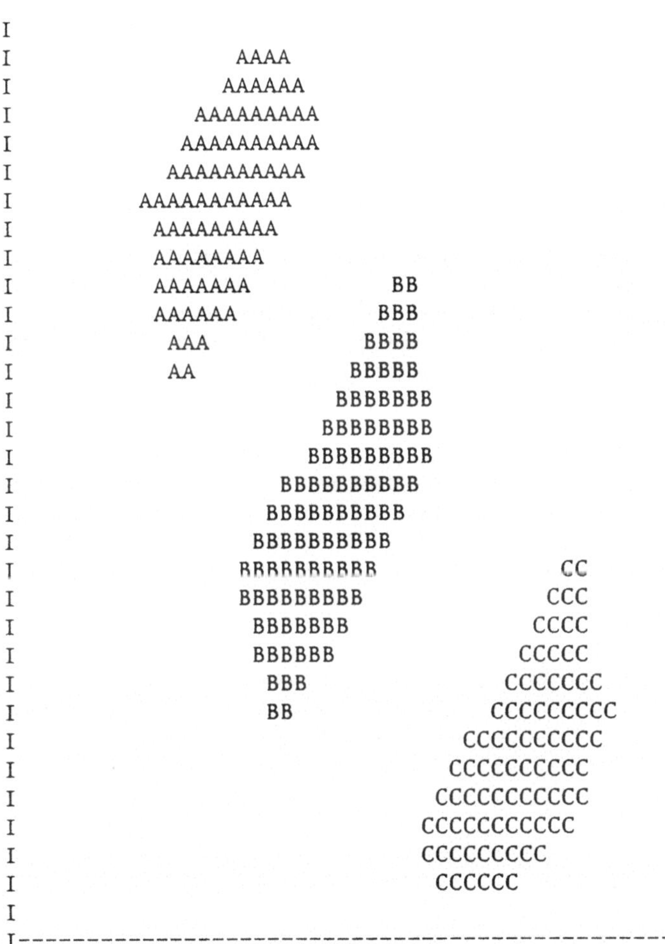

Figure 6
Positive Correlation within Clusters,
Negative Correlation Overall

which, since

$$\underline{T}'\underline{T} = \underline{\Sigma}^{-1},$$

is equal to

$$D^2(\underline{x1},\underline{x2};\ \underline{\Sigma}).$$

This is the square of the "Mahalanobis distance" between $\underline{x1}$ and $\underline{x2}$, in the metric of $\underline{\Sigma}$, where, in general, the square of the Mahalanobis distance between two vectors $\underline{v}$ and $\underline{w}$ in the metric of the nonsingular matrix $\underline{M}$ is denoted by $D^2(\underline{v},\underline{w};\ \underline{M})$ and is equal to the quadratic form

$$(\underline{v}-\underline{w})'\underline{M}^{-1}(\underline{v}-\underline{w}).$$

This shows that Mahalanobis distance is the appropriate measure of distance between observations on a random vector whose variables are correlated. Note that, given vectors $\underline{v}$ and $\underline{w}$,

$$D^2(\underline{v},\underline{w};\ \underline{I})$$

$$= \text{square of Euclidean distance between } \underline{v} \text{ and } \underline{w}.$$

As above, let $\underline{T}$ be a (nonsingular) matrix such that

$$\underline{T}\ \underline{\Sigma}\ \underline{T}' = \underline{I}.$$

Then

$$D^2(\underline{x1},\underline{x2};\ \underline{\Sigma})$$

$$= D^2(\underline{T}\ \underline{x1},\ \underline{T}\ \underline{x2};\ \underline{I})$$

$$= \text{square of Euclidean distance between } \underline{T}\ \underline{x1} \text{ and } \underline{T}\ \underline{x2}\ .$$

Thus, if we worked in terms of $\underline{T}\ \underline{x}$ instead of $\underline{x}$, Euclidean distance would be appropriate. Note that $\underline{T}$ depends upon $\underline{\Sigma}$, which has to be estimated from the sample. In cluster analysis, the observations are to be grouped, and $\underline{\Sigma}$ has to be estimated as a within-groups covariance matrix, not the total, overall covariance matrix.

Figure 6 depicts a situation where the within-cluster orientation is southwest-to-northeast but the overall orientation is northwest-to-southeast. That is, there is positive correlation within clusters, but negative correlation overall, in the mixed sample. Figure 6 shows what can happen if the overall, rather than the within-groups covariance matrix is used. The Mahalanobis distance would then erroneously adjust for negative rather than positive correlation. The importance of appropriate estimation of the covariance matrix in the clustering context will be expanded upon in the next section.

## 5. CLUSTERING: STATISTICAL INTERPRETATION OF "ISODATA"

In Sclove (1977) the ISODATA clustering procedure (Ball and Hall 1967) was studied. It was shown that, from the viewpoint of mathematical statistics, ISODATA corresponds to a method of iterated, maximum likelihood estimation in a mixture model for the clustering problem, where the distributions mixed are multivariate normal. The insight thus gained led to improvements in the algorithm, namely, using a within-groups covariance matrix, estimated as the clustering proceeds. This development will be reviewed here, and the ISODATA procedure as developed in Sclove (1977) will be discussed.

ISODATA proceeds as follows. Suppose two clusters are desired. [Later 3, 4, etc., can be tried. Model-selection criteria (see, e.g., Akaike 1983, 1985; Kashyap 1982) can be used as guides to the choice of the number k of clusters.] Choose two initial seed points as initial cluster centers. Loop through the dataset, assigning each individual to the seed point to which it is closest. Then, having tentatively assigned all n individuals, update the seed points, replacing them by the mean vectors of the tentatively formed groups. Then the individuals are reassigned, using the updated seed points. Etc. The procedure continues until no individual changes clusters.

If, rather than waiting for updating until a full pass is completed, one updates the seed point with the assignment of each individual, the resulting algorithm is MacQueen's (1966) k-means algorithm. (As indicated above, the symbol k is usually used to denote the number of clusters being formed.)

Now, as the ISODATA procedure was originally implemented (Ball and Hall 1967), it used Euclidean, or weighted Euclidean distance. But it can be seen that, once groups are tentatively formed, a tentative estimate of the within-groups covariance matrix can be made, and that can be used to compute Mahalanobis distances for clustering in the next pass. Thus, at each pass, the covariance matrix, as well as the seed points, can be updated.

The question raised in Sclove (1977) is whether there exists some model for the clustering problem such that ISODATA corresponds to an estimation scheme in the context of the model. The answer is that if one models the sample as having arisen from a mixture of multivariate normal distributions with different means, then ISODATA corresponds to a scheme for iterated maximum likelihood estimation in that model. The reason for this is understood by noting the connection between Mahalanobis distance and the multivariate normal probability density function. This connection is that the density function $f(\underline{x})$ depends upon $\underline{x}$ only through its Mahalanobis distance from the mean $\underline{\mu}$. That is,

$$f(\underline{x}) = \text{Const. } \exp[-\tfrac{1}{2}D^2(\underline{x},\underline{\mu};\ \Sigma)].$$

The conditional probability density function, given the c-th distribution, is

$$f(\underline{x}|c) = \text{Const. } \exp[-\tfrac{1}{2}D^2(\underline{x},\underline{\mu}_c;\ \Sigma)].$$

The seed points are initial estimates of the mean vectors. To maximize the likelihood in a given pass, one minimizes

$$D^2(\underline{x},\underline{\mu}_c;\ \Sigma).$$

That is, one assigns each observation $\underline{x}$ to that cluster c to whose mean it is closest, where the distance is measured by Mahalanobis distance in the metric of the covariance matrix. In each pass, one replaces the unknown parameters (mean vectors and covariance matrix) with their current estimates.

If the model is that of multivariate normal distributions with different covariance matrices, then

$$\log f(\underline{x}|c) = \text{Const.} - \tfrac{1}{2}\log|\Sigma_c| - \tfrac{1}{2}D^2(\underline{x},\underline{\mu}_c;\ \Sigma_c).$$

So $\underline{x}$ is not just clustered by minimum Mahalanobis distance; rather $\underline{x}$ is assigned to that cluster c for which

$$\log|\underline{\Sigma}_c| + D^2(\underline{x},\underline{\mu}_c;\ \underline{\Sigma}_c)$$

is minimal.

## 6. DISCUSSION; FUTURE RESEARCH

Marriott (1975) has pointed out that under the standard assumption of normal distributions with common covariance matrices, the maximization over possible labels (cluster assignments) gives inconsistent estimators for the parameters involved. Bryant and Williamson (1978) extended Marriott's results and showed that the method may be expected to give asymptotically biased results quite generally. See also McLachlan (1982).

The algorithms discussed in this paper are being modified to do estimation based on posterior probabilities of cluster membership rather than just maximizing over labels, i.e., rather than just assigning each observation to one cluster and estimating accordingly. It is expected that this will further improve the performance of the algorithms.

Department of Information and Decision Sciences
College of Business Administration m/c 294
University of Illinois at Chicago
Box 4348, Chicago, IL 60680-4348

## REFERENCES

The purposes of the various references below, with respect to this paper, are as follows. T.W. Anderson (1984) and Johnson and Wichern (1982) treat multivariate statistical analysis in general. The clustering algorithms mentioned in this paper are discussed in Ball and Hall (1967), MacQueen (1966), Sclove (1977), Solomon (1977), and Wolfe (1970). Sources of data mentioned in this paper include E. Anderson (1935), Fisher (1936), and Dixon and Massey (1969). The topic of metric considerations is the subject of Chernoff (1972). The book edited by Van Ryzin is the proceedings of an advanced seminar on classification and clustering. Questions related to the classification vs. the mixture-model likelihood are treated in Bryant and Williamson (1978), Marriott (1975) and McLachlan (1982). Some references on model-selection criteria are Akaike (1983, 1985) and Kashyap (1982).

Akaike,H. (1983). 'Statistical Inference and Measurement of Entropy.' In G.E.P. Box, T. Leonard, and C.-F. Wu (eds.), Scientific Inference, Data Analysis, and Robustness, 165-189. New York: Academic Press.

Akaike,H.(1985). 'Prediction and Entropy.' In A.C. Atkinson and S.E. Fienberg (eds.), A Celebration of Statistics: the ISI Centenary Volume, 1-24. New York: Springer-Verlag.

Anderson,E.(1935). 'The Irises of the Gaspe Peninsula,' Bulletin of the American Iris Society 59, 2-5.

Anderson,T.W.(1984). An Introduction to Multivariate Statistical Analysis, 2nd ed. New York: John Wiley and Sons.

Ball,G.H.,and Hall,D.J.(1967). 'A Clustering Technique for Summarizing Multivariate Data,' Behavioral Science 12, 153-155.

Bryant, P., and Williamson, J.A. (1978). 'Asymptotic Behavior of Classification Maximum Likelihood Estimates,' Biometrika 65, 273-281.

Chernoff, H. (1972). 'Metric Considerations in Cluster Analysis,' Proc. 6th Berkeley Symposium on Mathematical Statistics and Probability II, 621-630. Berkeley: University of California Press.

Dixon, W.J., and Massey, F.J. (1969). Introduction to Statistical Analysis, 3rd ed. New York: McGraw-Hill.

Fisher, R.A. (1936). 'The Use of Multiple Measurements in Taxonomic Problems,' Annals of Eugenics 7, 179-188.

Johnson, R.A., and Wichern, D.W. (1982). Applied Multivariate Statistical Analysis. New York: Prentice Hall.

Kashyap, R.L. (1982). 'Optimal Choice of AR and MA Parts in Autoregressive Moving Average Models, IEEE Transactions on Pattern Analysis and Machine Intelligence 4, 99-104.

MacQueen, J. (1966). 'Some Methods for Classification and Analysis of Multivariate Observations.' In Proc. 5th Berkeley Symposium on Mathematical Statistics and Probability I, 281-297. Berkeley: University of California Press.

McLachlan, G.J. (1982). 'The Classification and Mixture Maximum Likelihood Approaches to Cluster Analysis.' In P.R. Krishnaiah and L.N. Kanal (eds.), Handbook of Statistics 2 (Classification, Pattern Recognition and Reduction of Dimensionality), 199-208. New York: North Holland.

Marriott, F.H.C. (1975). 'Separating Mixtures of Normal Distributions,' Biometrics 31, 767-769.

Sclove, S.L. (1977). 'Population Mixture Models and Clustering Algorithms,' Communications in Statistics (A) 6, 417-434.

Solomon,H.(1977). 'Data Dependent Clustering Techniques,' In J. Van Ryzin (ed.), *Classification and Clustering*, 155-174. New York: Academic Press.

Van Ryzin,J.,ed.(1977). *Classification and Clustering*. New York: Academic Press.

Wolfe,J.H.(1970). 'Pattern Clustering by Multivariate Mixture Analysis,' *Multivariate Behavioral Research* *5*, 329-350.

# INDEX

The manufacturer's authorised representative in the EU is Springer Nature Customer Service Centre GmbH, Europaplatz 3, 69115 Heidelberg, Germany. If you have any concerns regarding our products, please contact ProductSafety@springernature.com

Printed and bound by CPI Group (UK) Ltd, Croydon, CR0 4YY
15/07/2026
02167619-0002